Fundamentals of Mathematical Logic

Fundamentals of Mathematical Logic

Samuel Parkers

STATES
ACADEMIC PRESS
www.statesacademicpress.com

Published by States Academic Press
109 South 5th Street,
Brooklyn, NY 11249, USA
www.statesacademicpress.com

Fundamentals of Mathematical Logic
Samuel Parkers

International Standard Book Number: 978-1-63989-228-0 (Hardback)

Cataloging-in-Publication Data

Fundamentals of mathematical logic / Samuel Parkers.
p. cm.
Includes bibliographical references and index.
ISBN 978-1-63989-228-0
1. Logic, Symbolic and mathematical. 2. Mathematics. 3. Logic. I. Parkers, Samuel.
QA9 .F86 2022
511.3--dc23

Table of Contents

Preface

It is with great pleasure that I present this book. It has been carefully written after numerous discussions with my peers and other practitioners of the field. I would like to take this opportunity to thank my family and friends who have been extremely supporting at every step in my life.

The sub-field of mathematics that focuses on identifying the applications of formal logic to mathematics is known as mathematical logic. It is also known as symbolic logic or formal logic. It is concerned with the study of expressive and deductive power of formal systems. Some of the formal logical systems are first-order logic, nonclassical and modal logic, algebraic logic and other classical logics. The discipline is divided into four areas. These are model theory, proof theory, set theory and recursion theory. The field is closely related to theoretical computer science and foundations of mathematics. The field finds its applications in other disciplines such as physics, biology, economics, metaphysics, law and morals, and psychology. This book explores all the important aspects of related to this discipline in the present day scenario. Different approaches, evaluations, methodologies and studies on mathematical logic have been included herein. As this field is emerging at a rapid pace, the contents of this book will help the readers understand the modern concepts and applications of the subject.

The chapters below are organized to facilitate a comprehensive understanding of the subject:

Chapter – What is Mathematical Logic?

The field of mathematics that applies formal logic to mathematics is termed as mathematical logic. Propositional logic, predicate logic and different rules of in-terference fall under its study. This chapter closely examines mathematical logic to provide an extensive understanding of the subject.

Chapter – Set Theory

Set theory is the branch of mathematical logic that studies well-determined collec-tion called sets. It includes combinatorial set theory, descriptive set theory, fuzzy set theory, inner model theory, Kripke-Platek theory, Venn diagram, cardinal func-tion, etc. This chapter has been carefully written to provide an easy understanding of the related concepts of set theory.

Chapter – Model Theory

The study of classes of groups, fields, graphs and other aspects of set theory under mathematical logic is termed as model theory. Finite model theory, first-order logic, Gödel's completeness and incompleteness theorem, etc. are studied within it. The top-ics elaborated in this chapter will help in gaining a better perspective of model theory.

Chapter – Proof Theory

Proof theory is the sub-field of mathematical logic which represents proofs as formal mathematical objects for their analysis by mathematical techniques. Some of its concepts are ordinal analysis, reverse mathematics, formal and informal proof, etc. This chapter delves into the concepts related to proof theory for a thorough understanding of it.

Chapter – Formal Logical Systems

Formal logical system refers to the set of inference rules that are used to conclude an expression, axioms and derived theorems. A few of its examples are axiomatic system, formal ethics, Lambda calculus, proof calculus, etc. This chapter sheds light on different formal logical systems to provide an in-depth understanding of the subject.

Samuel Parkers

1
What is Mathematical Logic?

The field of mathematics that applies formal logic to mathematics is termed as mathematical logic. Propositional logic, predicate logic and different rules of interference fall under its study. This chapter closely examines mathematical logic to provide an extensive understanding of the subject.

Mathematical logic is best understood as a branch of logic or mathematics. Mathematical logic is often divided into the subfields of model theory, proof theory, set theory and recursion theory. Research in mathematical logic has contributed to, and been motivated by, the study of foundations of mathematics, but mathematical logic also contains areas of pure mathematics not directly related to foundational questions.

One unifying theme in mathematical logic is the study of the expressive power of formal logics and formal proof systems. This power is measured both in terms of what these formal systems are able to prove and in terms of what they are able to define. Thus it can be said that "mathematical logic has become the general study of the logical structure of axiomatic theories."

Earlier names for mathematical logic were symbolic logic (as opposed to philosophical logic) and meta-mathematics. The former term is still used (as in the Association for Symbolic Logic), but the latter term is now used for certain aspects of proof theory.

Formal Logic

At its core, mathematical logic deals with mathematical concepts expressed using formal logical systems. The system of first-order logic is the most widely studied because of its applicability to foundations of mathematics and because of its desirable properties. Stronger classical logics such as second-order logic or infinitary logic are also studied, along with non-classical logics such as intuitionistic logic.

Fields of Mathematical Logic

Barwise's "Handbook of Mathematical Logic" divides mathematical logic into four parts:

- Set theory is the study of sets, which are abstract collections of objects. The basic concepts of set theory such as subset and relative complement are often called naive set theory. Modern research is in the area of axiomatic set theory, which uses logical methods to study which propositions are provable in various formal theories such as Zermelo-Frankel set theory, known as ZFC, or New Foundations set theory, known as NF.
- Proof theory is the study of formal proofs in various logical deduction systems. These proofs are represented as formal mathematical objects, facilitating their analysis by mathematical techniques. Frege worked on mathematical proofs and formalized the notion of a proof.
- Model theory studies the models of various formal theories. The set of all models of a particular theory is called an elementary class. Classical model theory seeks to determine the properties of models in a particular elementary class, or determine whether certain classes of structures form elementary classes. The method of quantifier elimination is used to show that models of particular theories cannot be too complicated.
- Recursion theory, also called computability theory, studies the properties of computable functions and the Turing degrees, which divide the uncomputable functions into sets which have the same level of uncomputability. The field has grown to include the study of generalized computability and definability. In these areas, recursion theory overlaps with proof theory and effective descriptive set theory.

The border lines between these fields, and also between mathematical logic and other fields of mathematics, are not always sharp; for example, Gödel's incompleteness theorem marks not only a milestone in recursion theory and proof theory, but has also led to Loeb's theorem, which is important in modal logic. The mathematical field of category theory uses many formal axiomatic methods resembling those used in mathematical logic, but category theory is not ordinarily considered a subfield of mathematical logic.

Connections with Computer Science

There are many connections between mathematical logic and computer science. Many early pioneers in computer science, such as Alan Turing, were also mathematicians and logicians.

The study of computability theory in computer science is closely related to the study of computability in mathematical logic. There is a difference of emphasis, however. Computer scientists often focus on concrete programming languages and feasible computability, while researchers in mathematical logic often focus on computability as a theoretical concept and on non-computability.

The study of programming language semantics is related to model theory, as is program verification (in particular, model checking). The Curry-Howard isomorphism between proofs and programs relates to proof theory; intuitionistic logic and linear logic are significant here. Calculi such as the lambda calculus and combinatory logic are nowadays studied mainly as idealized programming languages.

Computer science also contributes to mathematics by developing techniques for the automatic checking or even finding of proofs, such as automated theorem proving and logic programming.

Propositional Logic

A proposition is the basic building block of logic. It is defined as a declarative sentence that is either True or False, but not both.

The Truth Value of a proposition is True (denoted as T) if it is a true statement, and False (denoted as F) if it is a false statement.

For Example:

1. The sun rises in the East and sets in the West.

2. 1 + 1 = 2

3. 'b' is a vowel.

All of the above sentences are propositions, where the first two are Valid (True) and the third one is Invalid (False).

Some sentences that do not have a truth value or may have more than one truth value are not propositions.

For Example:

1. What time is it?

2. Go out and play.

3. x + 1 = 2.

The previous sentences are not propositions as the first two do not have a truth value, and the third one may be true or false.

To represent propositions, propositional variables are used. By Convention, these variables are represented by small alphabets such as p, q, r, s.

The area of logic which deals with propositions is called propositional calculus or propositional logic.

It also includes producing new propositions using existing ones. Propositions constructed using one or more propositions are called compound propositions. The propositions are combined together using Logical Connectives or Logical Operators.

Truth Table

Since we need to know the truth value of a proposition in all possible scenarios, we consider all the possible combinations of the propositions which are joined together by Logical Connectives to form the given compound proposition. This compilation of all possible scenarios in a tabular format is called a truth table.

Most Common Logical Connectives:

Negation: If p is a proposition, then the negation of p is denoted by ¬p, which when translated to simple English means- "It is not the case that p" or simply "not p". The truth value of ¬p is the opposite of the truth value of p.

The truth table of ¬p is:

P	¬p
T	F
F	T

Example: The negation of "It is raining today", is "It is not the case that is raining today" or simply "It is not raining today".

Conjunction: For any two propositions p and q, their conjunction is denoted by p ∧ q, which means "p and q". The conjuction p ∧ q is True when both p and q are True, otherwise False.

The truth table of p ∧ q is:

P	q	p ∧ q
T	T	T
T	F	F
F	T	F
F	F	F

Example: The conjunction of the propositions p – "Today is Friday" and q – "It is raining today", p ∧ q is "Today is Friday and it is raining today". This proposition is true only on rainy Fridays and is false on any other rainy day or on Fridays when it does not rain.

Disjunction: For any two propositions p and q, their disjunction is denoted by $p \vee q$, which means "p or q". The disjuction $p \vee q$ is True when either p or q is True, otherwise False.

The truth table of $p \vee q$ is:

P	q	$p \vee q$
T	T	T
T	F	T
F	T	T
F	F	F

Example: The disjunction of the propositions p – "Today is Friday" and q – "It is raining today", $p \vee q$ is "Today is Friday or it is raining today". This proposition is true on any day that is a Friday or a rainy day (including rainy Fridays) and is false on any day other than Friday when it also does not rain.

Exclusive Or: For any two propositions p and q, their exclusive or is denoted by $p \oplus q$, which means "either p or q but not both". The exclusive or $p \oplus q$ is True when either p or q is True, and False when both are true or both are false.

The truth table of $p \oplus q$ is:

P	q	$p \oplus q$
T	T	F
T	F	T
F	T	T
F	F	F

Example: The exclusive or of the propositions p – "Today is Friday" and q – "It is raining today", $p \oplus q$ is "Either today is Friday or it is raining today, but not both". This proposition is true on any day that is a Friday or a rainy day(not including rainy Fridays) and is false on any day other than Friday when it does not rain or rainy Fridays.

Implication: For any two propositions p and q, the statement "if p then q" is called an implication and it is denoted by $p \rightarrow q$.

In the implication $p \rightarrow q$, p is called the hypothesis or antecedent or premise and q is called the conclusion or consequence.

The implication is $p \rightarrow q$ is also called a conditional statement.

The implication is false when p is true and q is false otherwise it is true. The truth table of p $\rightarrow$ q is:

P	q	p $\rightarrow$ q
T	T	F
T	F	F
F	T	T
F	F	T

You might wonder that why is p $\rightarrow$ q true when p is false. This is because the implication guarantees that when p and q are true then the implication is true. But the implication does not guarantee anything when the premise p is false. There is no way of knowing whether or not the implication is false since p did not happen.

This situation is similar to the “Innocent until proven Guilty” stance, which means that the implication p $\rightarrow$ q is considered true until proven false. Since we cannot call the implication p $\rightarrow$ q false when p is false, our only alternative is to call it true.

This follows from the Explosion Principle which says:

“A False statement implies anything”.

Conditional statements play a very important role in mathematical reasoning, thus a variety of terminology is used to express p $\rightarrow$ q, some of which are listed below:

“if p, then q”

“p is sufficient for q”

“q when p”

“a necessary condition for p is q”

“p only if q”

“q unless ¬p”

“q follows from p”

Example: “If it is Friday then it is raining today” is a proposition which is of the form p $\rightarrow$ q. The above proposition is true if it is not Friday(premise is false) or if it is Friday and it is raining, and it is false when it is Friday but it is not raining.

Biconditional or Double Implication: For any two propositions p and q, the statement “p if and only if(iff) q” is called a biconditional and it is denoted by p $\leftrightarrow$ q.

The statement p $\leftrightarrow$ q is also called a bi-implication.

p $\leftrightarrow$ q has the same truth value as (p $\rightarrow$ q) $\wedge$ (q $\rightarrow$ p).

The implication is true when p and q have same truth values, and is false otherwise. The truth table of $p \leftrightarrow q$ is:

P	q	$p \leftrightarrow q$
T	T	T
T	F	F
F	T	F
F	F	T

Some other common ways of expressing $p \leftrightarrow q$ are:

"p is necessary and sufficient for q"

"if p then q, and conversely"

"p iff q"

Example: "It is raining today if and only if it is Friday today." is a proposition which is of the form $p \leftrightarrow q$. The previous proposition is true if it is not Friday and it is not raining or if it is Friday and it is raining, and it is false when it is not Friday or it is not raining.

De Morgan's Law

In propositional logic and boolean algebra, De Morgan's laws are a pair of transformation rules that are both valid rules of inference. They are named after Augustus De Morgan, a 19th-century British mathematician. The rules allow the expression of conjunctions and disjunctions purely in terms of each other via negation.

In formal language, the rules are written as:

- $\neg(p \wedge q) \equiv \neg p \vee \neg q$
- $\neg(p \vee q) \equiv \neg p \wedge \neg q$

Proof by Truth Table:

p	q	$\neg p$	$\neg q$	$p \wedge q$	$\neg p \vee \neg q$	$p \vee q$	$\neg p \wedge \neg q$
T	T	F	F	T	F	T	F
T	F	F	T	F	T	T	F
F	T	T	F	F	T	T	F
F	F	T	T	F	T	F	T

Special Conditional Statements:

As we know that we can form new propositions using existing propositions and logical connectives. New conditional statements can be formed starting with a conditional statement $p \rightarrow q$.

In particular, there are three related conditional statements that occur so often that they have special names.

- Implication: $p \rightarrow q$.
- Converse: The converse of the proposition $p \rightarrow q$ is $q \rightarrow p$.
- Contrapositive: The contrapositive of the proposition $p \rightarrow q$ is $\neg q \rightarrow \neg p$.
- Inverse: The inverse of the proposition $p \rightarrow q$ is $\neg p \rightarrow \neg q$.

To summarise,

Statement	If p, then q
Converse	If q, then p
Contrapositive	If not q, then not p
Inverse	If not p, then not q

It is interesting to note that the truth value of the conditional statement $p \rightarrow q$ is the same as it's contrapositive, and the truth value of the Converse of $p \rightarrow q$ is the same as the truth value of its Inverse.

When two compound propositions always have the same truth value, they are said to be equivalent.

Therefore,

- $p \rightarrow q \equiv \neg q \rightarrow \neg p$.
- $q \rightarrow p \equiv \neg p \rightarrow \neg q$.

p	q	$\neg p$	$\neg q$	$p \rightarrow q$	$\neg p \rightarrow \neg q$	$p \rightarrow q$	$\neg p \rightarrow \neg q$
T	T	F	F	T	T	T	T
T	F	F	T	F	F	T	T
F	T	T	F	T	T	F	F
F	F	T	T	T	T	T	T

Example:

Implication: If today is Friday, then it is raining.

The given proposition is of the form $p \rightarrow q$, where p is "Today is Friday" and q is "It is raining today".

Contrapositive, Converse, and Inverse of the given proposition respectively are:

- Converse: If it is raining, then today is Friday.

- Contrapositive: If it is not raining, then today is not Friday.
- Inverse: If today is not Friday, then it is not raining.

Implicit use of Biconditionals

In Natural Language bi-conditionals are not always explicit. In particular, the iff construction (if and only if) is rarely used in common language. Instead, bi-conditionals are often expressed using "if, then" or an "only if" construction. The other part of the "if and only if" is implicit, i.e. the converse is implied but not stated.

For example consider the following statement, "If you complete your homework, then you can go out and play". What is really meant is "You can go out and play if and only if you complete your homework". This statement is logically equivalent to two statements, "If you complete your homework, then you can go out and play" and "You can go out and play only if you complete your homework".

Because of this imprecision in Natural Language, an assumption needs to be made whether a conditional statement in natural language includes its converse or not.

Precedence Order of Logical Connectives

Logical connectives are used to construct compound propositions by joining existing propositions. Although parenthesis can be used to specify the order in which the logical operators in the compound proposition need to be applied, there exists a precedence order in Logical Operators.

The precedence Order is:

Operator	Precedence
$\neg$	1
$\wedge$	2
$\vee$	3
$\rightarrow$	4
$\leftrightarrow$	5

Here, higher the number lowers the precedence.

Translating English Sentences

Natural Languages such as English are ambiguous i.e. a statement may have multiple interpretations. Therefore it is important to convert these sentences into mathematical expressions involving propositional variables and logical connectives.

The above process of conversion may take certain reasonable assumptions about the intended meaning of the sentence. Once the sentences are translated into logical expressions they can be analyzed further to determine their truth values. Rules of Inference can then further be used to reason about the expressions.

Example: "You can access the Intenet from campus only if you are a computer science major or you are not a freshman."

The previous statement could be considered as a single proposition but it would be more useful to break it down into simpler propositions. That would make it easier to analyze its meaning and to reason with it.

The above sentence could be broken down into three propositions,

```
p - "You can access the Internet from campus."
q - "You are a computer science major."
r - "You are a freshman."
```

Using logical connectives we can join the mentioned propositions to get a logical expression of the given statement.

"Only if" is one way to express a conditional statement. Therefore the logical expression would be:

$$p \to (q \vee \neg r)$$

Predicate Logic

A predicate is an expression of one or more variables defined on some specific domain. A predicate with variables can be made a proposition by either assigning a value to the variable or by quantifying the variable.

The following are some examples of predicates:

- Let E(x, y) denote "x = y".
- Let X(a, b, c) denote "a + b + c = 0".
- Let M(x, y) denote "x is married to y".

Well formed Formula

Well Formed Formula (wff) is a predicate holding any of the following:

- All propositional constants and propositional variables are wffs.
- If x is a variable and Y is a wff, ∀xY and ∃xY are also wff.
- Truth value and false values are wffs.

- Each atomic formula is a wff.
- All connectives connecting wffs are wffs.

Quantifiers

The variable of predicates is quantified by quantifiers. There are two types of quantifier in predicate logic – Universal Quantifier and Existential Quantifier.

Universal Quantifier

Universal quantifier states that the statements within its scope are true for every value of the specific variable. It is denoted by the symbol $\forall$.

$\forall xP(x)$ is read as for every value of x, P(x) is true.

Example: "Man is mortal" can be transformed into the propositional form $\forall xP(x)$ where P(x) is the predicate which denotes x is mortal and the universe of discourse is all men.

Existential Quantifier

Existential quantifier states that the statements within its scope are true for some values of the specific variable. It is denoted by the symbol $\exists$.

$\exists x\ P(x)$ is read as for some values of x, P(x) is true.

Example: "Some people are dishonest" can be transformed into the propositional form $\exists xP(x)$ where P(x) is the predicate which denotes x is dishonest and the universe of discourse is some people.

Nested Quantifiers

If we use a quantifier that appears within the scope of another quantifier, it is called nested quantifier.

Example:

- $\forall\ a\ \exists\ bP(x,y)$ where P(a, b) denotes $a + b=0$
- $\forall\ a\ \forall\ b\ \forall\ cP(a, b, c)$ where P(a, b) denotes $a + (b + c) = (a + b) + c$

Note: $\forall\ a\ \exists\ bP(x, y) \neq \exists\ a\ \forall\ bP\ (x, y)$

Rules of Inference

To deduce new statements from the statements whose truth that we already know, Rules of Inference are used.

Mathematical logic is often used for logical proofs. Proofs are valid arguments that determine the truth values of mathematical statements.

An argument is a sequence of statements. The last statement is the conclusion and all its preceding statements are called premises (or hypothesis). The symbol "∴", (read therefore) is placed before the conclusion. A valid argument is one where the conclusion follows from the truth values of the premises.

Rules of Inference provide the templates or guidelines for constructing valid arguments from the statements that we already have.

Table: Rules of Inference.

Rule of Inference	Name	Rule of Inference	Name
P ——— $\therefore P \vee Q$	Addition	$P \vee Q$ $\neg P$ ——— $\therefore Q$	Disjunctive Syllogism
P Q ——— $\therefore P \wedge Q$	Conjunction	$P \rightarrow Q$ $Q \rightarrow R$ ——— $\therefore P \rightarrow R$	Hypothetical Syllogism
$P \wedge Q$ ——— $\therefore P$	Simplification	$(P \rightarrow Q) \wedge (R \rightarrow S)$ $P \vee R$ ——— $\therefore Q \vee S$	Constructive Dilemma
$P \rightarrow Q$ ———	Modus Ponens	$(P \rightarrow Q) \wedge (R \rightarrow S)$ $\neg Q \vee \neg S$ ——— $\therefore \neg P \vee \neg R$	Destructive Dilemma
$P \rightarrow Q$ $\neg Q$ ——— $\therefore \neg P$	Modus Tollens		

Addition

If P is a premise, we can use Addition rule to derive $P \vee Q$.

$$\frac{P}{\therefore P \vee Q}$$

Example:

Let P be the proposition, "He studies very hard" is true.

Therefore – "Either he studies very hard or he is a very bad student." Here Q is the proposition "he is a very bad student".

Conjunction

If P and Q are two premises, we can use Conjunction rule to derive $P \wedge Q$.

$$\begin{array}{c} P \\ Q \\ \hline \therefore P \wedge Q \end{array}$$

Example:

Let P – "He studies very hard".

Let Q – "He is the best boy in the class".

Therefore – "He studies very hard and he is the best boy in the class".

Simplification

If $P \wedge Q$ is a premise, we can use Simplification rule to derive P.

$$\begin{array}{c} P \wedge Q \\ \hline \therefore P \end{array}$$

Example:

"He studies very hard and he is the best boy in the class", $P \wedge Q$.

Therefore – "He studies very hard".

Modus Ponens

If P and $P \rightarrow Q$ are two premises, we can use Modus Ponens to derive Q.

$$\begin{array}{c} P \rightarrow Q \\ P \\ \hline \therefore Q \end{array}$$

Example:

"If you have a password, then you can log on to facebook", $P \rightarrow Q$.

"You have a password", P.

Therefore – "You can log on to facebook"

Modus Tollens

If $P \rightarrow Q$ and $\neg Q$ are two premises, we can use Modus Tollens to derive $\neg P$.

$$\begin{array}{l} P \rightarrow Q \\ \underline{\quad \neg Q} \\ \therefore \neg P \end{array}$$

Example:

"If you have a password, then you can log on to facebook", $P \rightarrow Q$.

"You cannot log on to facebook", $\neg Q$.

Therefore – "You do not have a password ".

Disjunctive Syllogism

If $\neg P$ and $P \vee Q$ are two premises, we can use Disjunctive Syllogism to derive Q.

$$\begin{array}{l} P \vee Q \\ \underline{\quad \neg P} \\ \therefore Q \end{array}$$

Example:

"The ice cream is not vanilla flavored", $\neg P$.

"The ice cream is either vanilla flavored or chocolate flavored", $P \vee Q$.

Therefore – "The ice cream is chocolate flavored".

Hypothetical Syllogism

If $P \rightarrow Q$ and $Q \rightarrow R$ are two premises, we can use Hypothetical Syllogism to derive $P \rightarrow R$.

$$\begin{array}{l} P \rightarrow Q \\ \underline{Q \rightarrow R} \\ \therefore P \rightarrow R \end{array}$$

Example:

"If it rains, I shall not go to school", $P \rightarrow Q$.

"If I don't go to school, I won't need to do homework", $Q \rightarrow R$.

Therefore – "If it rains, I won't need to do homework".

Constructive Dilemma

If $(P \rightarrow Q) \wedge (R \rightarrow S)$ and $P \vee R$ are two premises, we can use constructive dilemma to derive $Q \vee S$.

$$\begin{array}{c} (P \rightarrow Q) \wedge (R \rightarrow S) \\ P \vee R \\ \hline \therefore Q \vee S \end{array}$$

Example:

"If it rains, I will take a leave", $(P \rightarrow Q)$.

"If it is hot outside, I will go for a shower", $(R \rightarrow S)$.

"Either it will rain or it is hot outside", $P \vee R$.

Therefore – "I will take a leave or I will go for a shower".

Destructive Dilemma

If $(P \rightarrow Q) \wedge (R \rightarrow S)$ and $\neg Q \vee \neg S$ are two premises, we can use destructive dilemma to derive $\neg P \vee \neg R$.

$$\begin{array}{c} (P \rightarrow Q) \wedge (R \rightarrow S) \\ \neg Q \vee \neg S \\ \hline \therefore \neg P \vee \neg R \end{array}$$

Example:

"If it rains, I will take a leave", $(P \rightarrow Q)$.

"If it is hot outside, I will go for a shower", $(R \rightarrow S)$.

"Either I will not take a leave or I will not go for a shower", $\neg Q \vee \neg S$.

Therefore – "Either it does not rain or it is not hot outside".

References

- Mathematical-logic, entry: newworldencyclopedia.org, Retrieved 13 August, 2020
- Proposition-logic: geeksforgeeks.org, Retrieved 16 March, 2020
- Mathematical-logic-introduction-propositional-logic-set-2: geeksforgeeks.org, Retrieved 13 July, 2020
- Discrete-mathematics-predicate-logic, discrete-mathematics: tutorialspoint.com, Retrieved 29 May, 2020
- Rules-of-inference, discrete-mathematics: tutorialspoint.com, Retrieved 13 January, 2020

2
Set Theory

Set theory is the branch of mathematical logic that studies well-determined collection called sets. It includes combinatorial set theory, descriptive set theory, fuzzy set theory, inner model theory, Kripke-Platek theory, Venn diagram, cardinal function, etc. This chapter has been carefully written to provide an easy understanding of the related concepts of set theory.

Set Theory is a branch of mathematics that deals with the properties of well-defined collections of objects, which may or may not be of a mathematical nature, such as numbers or functions. The theory is less valuable in direct application to ordinary experience than as a basis for precise and adaptable terminology for the definition of complex and sophisticated mathematical concepts.

The theory had the revolutionary aspect of treating infinite sets as mathematical objects that are on an equal footing with those that can be constructed in a finite number of steps. Since antiquity, a majority of mathematicians had carefully avoided the introduction into their arguments of the actual infinite (i.e., of sets containing an infinity of objects conceived as existing simultaneously, at least in thought). Since this attitude persisted until almost the end of the 19th century, Cantor's work was the subject of much criticism to the effect that it dealt with fictions—indeed, that it encroached on the domain of philosophers and violated the principles of religion. Once applications to analysis began to be found, however, attitudes began to change, and by the 1890s Cantor's ideas and results were gaining acceptance. By 1900, set theory was recognized as a distinct branch of mathematics.

At just that time, however, several contradictions in so-called naive set theory were discovered. In order to eliminate such problems, an axiomatic basis was developed for the theory of sets analogous to that developed for elementary geometry. The degree of success that has been achieved in this development, as well as the present stature of set theory, has been well expressed in the Nicolas Bourbaki Éléments de mathématique: "Nowadays it is known to be possible, logically speaking, to derive practically the whole of known mathematics from a single source, The Theory of Sets."

Fundamental Set Concepts

In naive set theory, a set is a collection of objects (called members or elements) that is regarded as being a single object. To indicate that an object x is a member of a set A one writes $x \in A$, while $x \notin A$ indicates that x is not a member of A. A set may be defined by a membership rule (formula) or by listing its members within braces. For example, the set given by the rule "prime numbers less than 10" can also be given by {2, 3, 5, 7}. In principle, any finite set can be defined by an explicit list of its members, but specifying infinite sets requires a rule or pattern to indicate membership; for example, the ellipsis in {0, 1, 2, 3, 4, 5, 6, 7, ..., etc.} indicates that the list of natural numbers N goes on forever. The empty (or void, or null) set, symbolized by {} or Ø, contains no elements at all. Nonetheless, it has the status of being a set.

A set A is called a subset of a set B (symbolized by $A \subseteq B$) if all the members of A are also members of B. For example, any set is a subset of itself, and Ø is a subset of any set. If both $A \subseteq B$ and $B \subseteq A$, then A and B have exactly the same members. Part of the set concept is that in this case A = B; that is, A and B are the same set.

Relations in Set Theory

In mathematics, a relation is an association between, or property of, various objects. Relations can be represented by sets of ordered pairs (a, b) where a bears a relation to b. Sets of ordered pairs are commonly used to represent relations depicted on charts and graphs, on which, for example, calendar years may be paired with automobile production figures, weeks with stock market averages, and days with average temperatures.

A function f can be regarded as a relation between each object x in its domain and the value f(x). A function f is a relation with a special property, however: each x is related by f to one and only one y. That is, two ordered pairs (x, y) and (x, z) in f imply that y = z.

A one-to-one correspondence between sets A and B is similarly a pairing of each object in A with one and only one object in B, with the dual property that each object in B has been thereby paired with one and only one object in A. For example, if A = {x, z, w} and B = {4, 3, 9}, a one-to-one correspondence can be obtained by pairing x with 4, z with 3, and w with 9. This pairing can be represented by the set {(x, 4), (z, 3), (w, 9)} of ordered pairs.

Many relations display identifiable properties. For example, in the relation "is the same colour as," each object bears the relation to itself as well as to some other objects. Such relations are said to be reflexive. The ordering relation "less than or equal to" (symbolized by $\leq$) is reflexive, but "less than" (symbolized by $<$) is not. The relation "is parallel to" (symbolized by $\parallel$) has the property that, if an object bears the relation to a second object, then the second also bears that relation to the first. Relations with this property are said to be symmetric. (Note that the ordering relation is not symmetric.) These examples also have the property that whenever one object bears the relation to a second, which further bears the relation to a third, then the first bears that relation to the third—e.g., if a < b and b < c, then a < c. Such relations are said to be transitive.

Relations that have all three of these properties—reflexivity, symmetry, and transitivity—are called equivalence relations. In an equivalence relation, all elements related to a particular element, say a, are also related to each other, and they form what is called the equivalence class of a. For example, the equivalence class of a line for the relation "is parallel to" consists of the set of all lines parallel to it.

Essential Features of Cantorian Set Theory

At best, the foregoing description presents only an intuitive concept of a set. Essential features of the concept as Cantor understood it include: (1) that a set is a grouping into a single entity of objects of any kind, and (2) that, given an object x and a set A, exactly one of the statements $x \in A$ and $x \notin A$ is true and the other is false. The definite relation that may or may not exist between an object and a set is called the membership relation.

A further intent of this description is conveyed by what is called the principle of extension—a set is determined by its members rather than by any particular way of describing the set. Thus, sets A and B are equal if and only if every element in A is also in B and every element in B is in A; symbolically, $x \in A$ implies $x \in B$ and vice versa. There exists, for example, exactly one set the members of which are 2, 3, 5, and 7. It does not matter whether its members are described as "prime numbers less than 10" or listed in some order (which order is immaterial) between small braces, possibly {5, 2, 7, 3}.

The positive integers {1, 2, 3, ..., etc.} are typically used for counting the elements in a finite set. For example, the set {a, b, c} can be put in one-to-one correspondence with the elements of the set {1, 2, 3}. The number 3 is called the cardinal number, or cardinality, of the set {1, 2, 3} as well as any set that can be put into a one-to-one correspondence with it. (Because the empty set has no elements, its cardinality is defined as 0.) In general, a set A is finite and its cardinality is n if there exists a pairing of its elements with the set {1, 2, 3, ..., n}. A set for which there is no such correspondence is said to be infinite.

To define infinite sets, Cantor used predicate formulas. The phrase "x is a professor" is an example of a formula; if the symbol x in this phrase is replaced by the name of a person; there results a declarative sentence that is true or false. The notation S(x) will be used to represent such a formula. The phrase "x is a professor at university y and x is a male" is a formula with two variables. If the occurrences of x and y are replaced by names of appropriate, specific objects, the result is a declarative sentence that is true or false. Given any formula S(x) that contains the letter x (and possibly others), Cantor's principle of abstraction asserts the existence of a set A such that, for each object x, $x \in A$ if and only if S(x) holds. (Mathematicians later formulated a restricted principle of abstraction, also known as the principle of comprehension, in which self-referencing predicates, or S(A), are excluded in order to prevent certain paradoxes. Cardinality and transfinite numbers.) Because of the principle of extension, the set A corresponding to S(x) must be unique, and it is symbolized by {x | S(x)}, which is read "The set of all objects x such that S(x)." For instance, {x | x is blue} is the set of all blue objects. This illustrates the fact that the principle of abstraction implies the existence of sets the elements of which are all objects having a certain property. It is actually more comprehensive. For example, it asserts the existence of a set B corresponding to "Either x is an astronaut or x is a natural number." Astronauts have no particular property in common with numbers (other than both being members of B).

Equivalent Sets

Cantorian set theory is founded on the principles of extension and abstraction. To describe some results based upon these principles, the notion of equivalence of sets will be defined. The idea is that two sets are equivalent if it is possible to pair off members of the first set with members of the second, with no leftover members on either side. To capture this idea in set-theoretic terms, the set A is defined as equivalent to the set B (symbolized by A ≡ B) if and only if there exists a third set the members of which are ordered pairs such that: (1) the first member of each pair is an element of A and the second is an element of B, and (2) each member of A occurs as a first member and each member of B occurs as a second member of exactly one pair. Thus, if A and B are finite and A ≡ B, then the third set that establishes this fact provides a pairing, or matching, of the elements of A with those of B. Conversely, if it is possible to match the elements of A with those of B, then A ≡ B, because a set of pairs meeting requirements (1) and (2) can be formed—i.e., if a ∈ A is matched with b ∈ B, then the ordered pair (a, b) is one member of the set. By thus defining equivalence of sets in terms of the notion of matching, equivalence is formulated independently of finiteness. As an illustration involving infinite sets, N may be taken to denote the set of natural numbers 0, 1, 2, ..., etc. Then {(n, n2) | n ∈ N} establishes the seemingly paradoxical equivalence of N and the subset of N formed by the squares of the natural numbers.

A set B is included in, or is a subset of, a set A (symbolized by B ⊆ A) if every element of B is an element of A. So defined, a subset may possibly include all of the elements of A, so that A can be a subset of itself. Furthermore, the empty set, because it by definition has no elements that are not included in other sets, is a subset of every set.

If every element of set B is an element of set A, but the converse is false (hence B ≠ A), then B is said to be properly included in, or is a proper subset of, A (symbolized by B ⊂ A). Thus, if A = {3, 1, 0, 4, 2}, both {0, 1, 2} and {0, 1, 2, 3, 4} are subsets of A; but {0, 1, 2, 3, 4} is not a proper subset. A finite set is nonequivalent to each of its proper subsets. This is not so, however, for infinite sets, as is illustrated with the set N in the earlier example. (The equivalence of N and its proper subset formed by the squares of its elements was noted by Galileo Galilei in 1638, who concluded that the notions of less than, equal to, and greater than did not apply to infinite sets.)

Cardinality and Transfinite Numbers

The application of the notion of equivalence to infinite sets was first systematically explored by Cantor. With N defined as the set of natural numbers, Cantor's initial significant finding was that the set of all rational numbers is equivalent to N but that the set of all real numbers is not equivalent to N. The existence of nonequivalent infinite sets justified Cantor's introduction of "transfinite" cardinal numbers as measures of size for such sets. Cantor defined the cardinal of an arbitrary set A as the concept that can be abstracted from A taken together with the totality of other equivalent sets. Gottlob Frege, in 1884, and Bertrand Russell, in 1902, both mathematical logicians, defined the cardinal number Depiction of the cardinal number $\overline{\overline{A}}$ of a set A somewhat more explicitly, as the set of all sets that are equivalent to A. This definition thus provides a place for cardinal numbers as objects of a universe whose only members are sets.

The previous definitions are consistent with the usage of natural numbers as cardinals. Intuitively, a cardinal number, whether finite (i.e., a natural number) or transfinite (i.e., nonfinite), is a measure of the size of a set. Exactly how a cardinal number is defined is unimportant; what is important is that $\overline{\overline{A}} = \overline{\overline{B}}$ if and only if $A \equiv B$.

To compare cardinal numbers, an ordering relation (symbolized by <) may be introduced by means of the definition $\overline{\overline{A}} < \overline{\overline{B}}$ if A is equivalent to a subset of B and B is equivalent to no subset of A. Clearly, this relation is irreflexive $\overline{\overline{A}} \not< \overline{\overline{A}}$ and transitive: $\overline{\overline{A}} < \overline{\overline{B}}$ and $\overline{\overline{B}} < \overline{\overline{C}}$ imply $\overline{\overline{A}} < \overline{\overline{C}}$.

When applied to natural numbers used as cardinals, the relation < (less than) coincides with the familiar ordering relation for N, so that < is an extension of that relation.

The symbol ℵo (aleph-null) is standard for the cardinal number of N (sets of this cardinality are called denumerable), and ℵ (aleph) is sometimes used for that of the set of real numbers. Then $n < \aleph_0$ for each $n \in N$ and $\aleph_0 < \aleph$.

This, however, is not the end of the matter. If the power set of a set A—symbolized P(A)—is defined as the set of all subsets of A, then, as Cantor proved, $\overline{\overline{A}} < \overline{\overline{P(A)}}$ for every set A – a relation that is known as Cantor's theorem. It implies an unending hierarchy of transfinite cardinals: $\overline{\overline{N}} = \aleph_0 \overline{\overline{P(N)}}, \overline{\overline{P(P(N))}}, \ldots,$ etc. Cantor proved that $\aleph = \overline{\overline{P(N)}}$ and suggested that there are no cardinal numbers between $\aleph_0$ and ℵ, a conjecture known as the continuum hypothesis.

There is an arithmetic for cardinal numbers based on natural definitions of addition, multiplication, and exponentiation (squaring, cubing, and so on), but this arithmetic deviates from that of the natural numbers when transfinite cardinals are involved. For example, $\aleph_0 + \aleph_0 = \aleph_0$ (because the set of integers is equivalent to N), $\aleph_0 \cdot \aleph_0 = \aleph_0$ (because the set of ordered pairs of natural numbers is equivalent to N), and $c + \aleph_0 = c$ for every transfinite cardinal c (because every infinite set includes a subset equivalent to N).

The so-called Cantor paradox, discovered by Cantor himself in 1899, is the following. By the unrestricted principle of abstraction, the formula "x is a set" defines a set U; i.e., it is the set of all sets. Now P(U) is a set of sets and so P(U) is a subset of U. By the definition of < for cardinals, however, if $A \subseteq B$, then it is not the case that $\overline{\overline{B}} < \overline{\overline{A}}$. Hence, by substitution, $\overline{\overline{U}} \not< \overline{\overline{P(U)}}$. But by Cantor's theorem, $\overline{\overline{U}} < \overline{\overline{P(U)}}$. This is a contradiction. In 1901 Russell devised another contradiction of a less technical nature that is now known as Russell's paradox. The formula "x is a set and $(x \notin x)$" defines a set R of all sets not members of themselves. Using proof by contradiction, however, it is easily shown that (1) $R \in R$. But then by the definition of R it follows that (2) $(R \notin R)$. Together, (1) and (2) form a contradiction.

Axiomatic Set Theory

In contrast to naive set theory, the attitude adopted in an axiomatic development of set theory is that it is not necessary to know what the "things" are that are called "sets" or what the relation of membership means. Of sole concern are the properties assumed about sets and the membership relation. Thus, in an axiomatic theory of sets, set and the membership relation $\in$ are undefined terms. The assumptions adopted about these notions are called the axioms of the theory. Axiomatic set theorems are the axioms together with statements that can be deduced from the axioms using the rules of inference provided by a system of logic. Criteria for the choice of axioms include: (1) consistency—it should be impossible to derive as theorems both a statement and its negation; (2) plausibility—axioms should be in accord with intuitive beliefs about sets; and (3) richness—desirable results of Cantorian set theory can be derived as theorems.

Schemas for Generating Well-formed Formulas

The ZFC "axiom of extension" conveys the idea that, as in naive set theory, a set is determined solely by its members. It should be noted that this is not merely a logically necessary property of equality but an assumption about the membership relation as well.

The set defined by the "axiom of the empty set" is the empty (or null) set Ø.

For an understanding of the "axiom schema of separation" considerable explanation is required. Zermelo's original system included the assumption that, if a formula S(x) is "definite" for all elements of a set A, then there exists a set the elements of which are precisely those elements x of A for which S(x) holds. This is a restricted version of the principle of abstraction, now known as the principle of comprehension, for it provides for the existence of sets corresponding to formulas. It restricts that principle, however, in two ways: (1) Instead of asserting the existence of sets unconditionally, it can be applied only in conjunction with preexisting sets, and (2) only "definite" formulas may be used. Zermelo offered only a vague description of "definite," but clarification was given by Skolem by way of a precise definition of what will be called simply a formula of ZFC. Using tools of modern logic, the definition may be made as follows:

- For any variables x and y, x ∈ y and x = y are formulas (such formulas are called atomic).
- If S and T are formulas and x is any variable, then each of the following is a formula: If S, then T; S if and only if T; S and T; S or T; not S; for all x, S; for some x, T.

Formulas are constructed recursively (in a finite number of systematic steps) beginning with the (atomic) formulas of (I) and proceeding via the constructions permitted in (II). "Not (x ∈ y)," for example, is a formula (which is abbreviated to x ∉ y), and "There exists an x such that for every y, y ∉ x" is a formula. A variable is free in a formula if it occurs at least once in the formula without being introduced by one of the phrases "for some x" or "for all x." Henceforth, a formula S in which x occurs as a free variable will be called "a condition on x" and symbolized S(x). The formula "For every y, x ∈ y," for example, is a condition on x. It is to be understood that a formula is a formal expression–i.e., a term without meaning. Indeed, a computer can be programmed to generate atomic formulas and build up from them other formulas of ever-increasing complexity using logical connectives ("not," "and," etc.) and operators ("for all" and "for some"). A formula acquires meaning only when an interpretation of the theory is specified; i.e., when (1) a nonempty collection (called the domain of the interpretation) is specified as the range of values of the variables (thus the term set is assigned a meaning, viz., an object in the domain), (2) the membership relation is defined for these sets, (3) the logical connectives and operators are interpreted as in everyday language, and (4) the logical relation of equality is taken to be identity among the objects in the domain.

The phrase "a condition on x" for a formula in which x is free is merely suggestive; relative to an interpretation, such a formula does impose a condition on x. Thus, the intuitive interpretation of the "axiom schema of separation" is: given a set A and a condition on x, S(x), those elements of A for which the condition holds form a set. It provides for the existence of sets by separating off certain elements of existing sets. Calling this the axiom schema of separation is appropriate, because it is actually a schema for generating axioms—one for each choice of S(x).

Axioms for Compounding Sets

Although the axiom schema of separation has a constructive quality, further means of constructing sets from existing sets must be introduced if some of the desirable features of Cantorian set theory are to be established. Three axioms in the table–axiom of pairing, axiom of union, and axiom of power set–are of this sort.

By using five of the axioms, a variety of basic concepts of naive set theory (e.g., the operations of union, intersection, and Cartesian product; the notions of relation, equivalence relation, ordering relation, and function) can be defined with ZFC. Further, the standard results about these concepts that were attainable in naive set theory can be proved as theorems of ZFC.

Axioms for Infinite and Ordered Sets

If I is an interpretation of an axiomatic theory of sets, the sentence that results from an axiom when a meaning has been assigned to "set" and "∊," as specified by I, is either true or false. If each axiom is true for I, then I is called a model of the theory. If the domain of a model is infinite, this fact does not imply that any object of the domain is an "infinite set." An infinite set in the latter sense is an object d of the domain D of I for which there is infinity of distinct objects d′ in D such that d′Ed holds (E standing for the interpretation of ∊). Though the domain of any model of the theory of which the axioms thus far discussed are axioms is clearly infinite, models in which every set is finite have been devised. For the full development of classical set theory, including the theories of real numbers and of infinite cardinal numbers, the existence of infinite sets is needed; thus the "axiom of infinity" is included.

The existence of a unique minimal set ω having properties expressed in the axiom of infinity can be proved; its distinct members are Ø, {Ø}, {Ø, {Ø}}, {Ø, {Ø}, {Ø, {Ø}}}, ..., etc. These elements are denoted by 0, 1, 2, 3, ..., etc. and are called natural numbers. Justification for this terminology rests with the fact that the Peano postulates (five axioms published in 1889 by the Italian mathematician Giuseppe Peano), which can serve as a base for arithmetic, can be proved as theorems in set theory. Thereby the way is paved for the construction within ZFC of entities that have all the expected properties of the real numbers.

The origin of the axiom of choice was Cantor's recognition of the importance of being able to "well-order" arbitrary sets—i.e., to define an ordering relation for a given set such that each nonempty subset has a least element. The virtue of a well-ordering for a set is that it offers a means of proving that a property holds for each of its elements by a process (transfinite induction) similar to mathematical induction. Zermelo gave the first proof that any set can be well-ordered. His proof employed a set-theoretic principle that he called the "axiom of choice," which, shortly thereafter, was shown to be equivalent to the so-called well-ordering theorem.

Intuitively, the axiom of choice asserts the possibility of making a simultaneous choice of an element in every nonempty member of any set; this guarantee accounts for its name. The assumption is significant only when the set has infinitely many members. Zermelo was the first to state explicitly the axiom, although it had been used but essentially unnoticed earlier. It soon became the subject of vigorous controversy because of its nonconstructive nature. Some mathematicians rejected it totally on this ground.

Others accepted it but avoided its use whenever possible. Some changed their minds about it when its equivalence with the well-ordering theorem was proved as well as the assertion that any two cardinal numbers c and d are comparable (i.e., that exactly one of $c < d$, $d < c$, $c = d$ holds). There are many other equivalent statements, though even today a few mathematicians feel that the use of the axiom of choice is improper. To the vast majority, however, it, or an equivalent assertion, has become an indispensable and commonplace tool. (Because of this controversy, ZFC was adopted as an acronym for the majority position with the axiom of choice and ZF for the minority position without the axiom of choice.)

Schema for Transfinite Induction and Ordinal Arithmetic

When Zermelo's axioms 1–8 were found to be inadequate for a full-blown development of transfinite induction and ordinal arithmetic, Fraenkel and Skolem independently proposed an additional axiom schema to eliminate the difficulty. As modified by Hungarian-born American mathematician John von Neumann, it says, intuitively, that if with each element of a set there is associated exactly one set, then the collection of the associated sets is itself a set; i.e., it offers a way to "collect" existing sets to form sets. As an illustration, each of ω, $P(\omega)$, $P(P(\omega))$, ..., etc. formed by recursively taking power sets (set formed of all the subsets of the preceding set), is a set in the theory based on Zermelo's original eight axioms. But there appears to be no way to establish the existence of the set having all these sets as its members. However, an instance of the "axiom schema of replacement" provides for its existence.

Intuitively, the axiom schema of replacement is the assertion that, if the domain of a function is a set, then so is its range. That this is a powerful schema (in respect to the further inferences that it yields) is suggested by the fact that the axiom schema of separation can be derived from it and that, when applied in conjunction with the axiom of power set, the axiom of pairing can be deduced.

The axiom schema of replacement has played a significant role in developing a theory of ordinal numbers. In contrast to cardinal numbers, which serve to designate the size of a set, ordinal numbers are used to determine positions within a prescribed well-ordered sequence. Under an approach conceived by von Neumann, if A is a set, the successor A′ of A is the set obtained by adjoining A to the elements of A ($A' = A \cup \{A\}$). In terms of this notion the natural numbers, as defined above, are simply the succession $0, 0', 0'', 0'''$, ..., etc.; i.e., the natural numbers are the sets obtained starting with Ø and iterating the prime operation a finite number of times. The natural numbers are well-ordered by the $\in$ relation, and with this ordering they constitute the finite ordinal numbers. The axiom of infinity secures the existence of the set of natural numbers, and the set ω is the first infinite ordinal. Greater ordinal numbers are obtained by iterating the prime operation beginning with ω. An instance of the axiom schema of replacement asserts that $\omega, \omega', \omega''$, ..., etc. form a set. The union of this set and ω is the still greater ordinal that is denoted by $\omega 2$ (employing notation from ordinal arithmetic). A repetition of this process beginning with ω2 yields the ordinals (ω2)′, (ω2)″, ..., etc.; next after all of those of this form is ω3. In this way the sequence of ordinals ω, ω2, ω3, ..., etc. is generated. An application of the axiom schema of replacement then yields the ordinal that follows all of these in the same sense in which ω follows the finite ordinals; in the notation from ordinal arithmetic, it is ω^2. At this point the iteration process can be repeated. In summary, the axiom schema of replacement together with the other axioms makes possible the extension of the counting process as far beyond the natural numbers as one chooses.

In the ZFC system, cardinal numbers are defined as certain ordinals. From the well-ordering theorem (a consequence of the axiom of choice), it follows that every set A is equivalent to some ordinal number. Also, the totality of ordinals equivalent to A can be shown to form a set. Then a natural choice for the cardinal number of A is the least ordinal to which A is equivalent. This is the motivation for defining a cardinal number as an ordinal that is not equivalent to any smaller ordinal. The arithmetics of both cardinal and ordinal numbers have been fully developed. That of finite cardinals and ordinals coincides with the arithmetic of the natural numbers. For infinite cardinals, the arithmetic is uninteresting since, as a consequence of the axiom of choice, both the sum and product of two such cardinals are equal to the maximum of the two. In contrast, the arithmetic of infinite ordinals is interesting and presents a wide assortment of oddities.

In addition to the guidelines already mentioned for the choice of axioms of ZFC, another guideline is taken into account by some set theorists. For the purposes of foundational studies of mathematics, it is assumed that mathematics is consistent; otherwise, any foundation would fail. It may thus be reasoned that, if a precise account of the intuitive usages of sets by mathematicians is given, an adequate and correct foundation will result. Traditionally, mathematicians deal with the integers, with real numbers, and with functions. Thus, an intuitive hierarchy of sets in which these entities appear should be a model of ZFC. It is possible to construct such a hierarchy explicitly from the empty set by iterating the operations of forming power sets and unions in the following way.

The bottom of the hierarchy is composed of the sets $A_0 = \emptyset$, A_1, ..., A_n, ..., etc. in which each A_{n+1} is the power set of the preceding An. Then one can form the union A_ω of all sets constructed thus far. This can be followed by iterating the power set operation as before: $A_{\omega'}$ is the power set of A_ω and so forth. This construction can be extended to arbitrarily high transfinite levels. There is no highest level of the hierarchy; at each level, the union of what has been constructed thus far can be taken and the power set operation applied to the elements. In general, for each ordinal number α one obtains a set A_α, each member of which is a subset of some A_β that is lower in the hierarchy. The hierarchy obtained in this way is called the iterative hierarchy. The domain of the intuitive model of ZFC is conceived as the union of all sets in the iterative hierarchy. In other words, a set is in the model if it is an element of some set A_α of the iterative hierarchy.

Axiom for Eliminating Infinite Descending Species

From the assumptions that this system of set theory is sufficiently comprehensive for mathematics and that it is the model to be "captured" by the axioms of ZFC, it may be argued that models of axioms 1 through 9 of the table that differ sharply from this system should be ruled out. The discovery of such a model led to the formulation by von Neumann of axiom 10, the axiom of restriction, or foundation axiom.

This axiom eliminates from the models of the first nine axioms those in which there exist infinite descending $\in$-chains (i.e., sequences $x_1, x_2, x_3, \ldots$, etc. such that $x_2 \in x_1$, $x_3 \in x_2, \ldots$, etc.), a phenomenon that does not appear in the model based on an iterative hierarchy. It also has other attractive consequences; e.g., a simpler definition of the notion of ordinal number is possible. Yet there is no unanimity among mathematicians whether there are sufficient grounds for adopting it as an additional axiom. On the one hand, the axiom is equivalent (in a theory that allows only sets) to the statement that every set appears in the iterative hierarchy informally described above—there are no other sets. So it formulates the view that this is what the universe of all sets is really like. On the other hand, there is no compelling need to rule out sets that might lie outside the hierarchy—the axiom has not been shown to have any mathematical applications.

The Neumann-Bernays-Gödel Axioms

Neumann-Bernays-Gödel axioms

(1) *Axiom of extension.* If A and B are classes and if, for all (sets) x, $x \in A$ if and only if $x \in B$, then $A = B$.

(2) *Axiom of the empty set.* There exists a set A such that, for all x, it is false that $x \in A$.

(3) *Axiom schema for class formation.* If $S(x)$ is a condition on x in which only set variables are introduced by the phrase "for all" or "for some" and in which B is not free, then there exists a class B such that $x \in B$ if and only if $S(x)$.

(4) *Axiom of pairing.* If A and B are sets, there exists a set (symbolized $\{A, B\}$ and called the unordered pair of A and B) having A and B as its sole members.

(5) *Axiom of union.* If C is a set, there exists a set A such that $x \in A$ if and only if $x \in B$ for some member B of C.

(6) *Axiom of power set.* If A is a set, there exists a set B, called its power set, such that $x \in B$ if and only if $x \subseteq A$.

(7) *Axiom of infinity.* There exists a set A such that $\varnothing \in A$ and, if $x \in A$, then $(x \cup \{x\}) \in A$, in which $x \cup \{x\}$ is the set x with x adjoined as a further member.

(8) *Axiom of choice.* If A is a set the elements of which are nonempty sets, then there exists a function f with domain A such that, for each member B of A, $f(B) \in B$.

(9) *Axiom of replacement.* If (the class) X is a function and A is a set, then there exists a set B such that $y \in B$ if and only if, for some x, $(x, y) \in X$ and $x \in A$; i.e., the range of the restriction of a function X to a domain that is a set is also a set.

(10) *Axiom of restriction (foundation axiom).* Every nonempty class A contains an element B such that $A \cap B = \varnothing$.

The second axiomatization of set theory originated with John von Neumann in the 1920s. His formulation differed considerably from ZFC because the notion of function, rather than that of set, was taken as undefined, or "primitive." In a series of papers beginning in 1937, however, the Swiss logician Paul Bernays, a collaborator with the German formalist David Hilbert, modified the von Neumann approach in a way that put it in much closer contact with ZFC. In 1940, the Austrian-born American logician Kurt Gödel, known for his undecidability proof, further simplified the theory. This axiomatic version of set theory is called NBG, after the Neumann-Bernays-Gödel axioms. As will be explained shortly, NBG is closely related to ZFC, but it allows explicit treatment of so-called classes: collections that might be too large to be sets, such as the class of all sets or the class of all ordinal numbers.

For expository purposes it is convenient to adopt two undefined notions for NBG: class and the binary relation $\in$ of membership (though, as is also true in ZFC, $\in$ suffices). For the intended interpretation, variables take classes—the totalities corresponding to certain properties—as values. A class is defined to be a set if it is a member of some class; those classes that are not sets are called proper classes. Intuitively, sets are intended to be those classes that are adequate for mathematics, and proper classes are thought of as those collections that are "so big" that, if they were permitted to be sets, contradictions would follow. In NBG, the classical paradoxes are avoided by proving in each case that the collection on which the paradox is based is a proper class—i.e., is not a set.

Comments about the axioms that follow are limited to features that distinguish them from their counterpart in ZFC. The axiom schema for class formation is presented in a form to facilitate a comparison with the axiom schema of separation of ZFC. In a detailed development of NBG, however, there appears instead a list of seven axioms (not schemas) that state that, for each of certain conditions, there exists a corresponding class of all those sets satisfying the condition. From this finite set of axioms, each an instance of the above schema, the schema (in a generalized form) can be obtained as a theorem. When obtained in this way, the axiom schema for class formation of NBG is called the class existence theorem.

In brief, axioms 4 through 8 in the table of NBG are axioms of set existence. The same is true of the next axiom, which for technical reasons is usually phrased in a more general form. Finally, there may appear in a formulation of NBG an analog of the last axiom of ZFC (axiom of restriction).

A comparison of the two theories that have been formulated is in order. In contrast to the axiom schema of replacement of ZFC, the NBG version is not an axiom schema but an axiom. Thus, with the comments above about the ZFC axiom schema of separation in mind, it follows that NBG has only a finite number of axioms. On the other hand, since the axiom schema of replacement of ZFC provides an axiom for each formula, ZFC has infinitely many axioms—which is unavoidable because it is known that no finite subset yields the full system of axioms. The finiteness of the axioms for NBG makes the logical study of the system simpler. The relationship between the theories may be summarized by the statement that ZFC is essentially the part of NBG that refers only to sets. Indeed, it has been proved that every theorem of ZFC is a theorem of NBG and that any theorem of NBG that speaks only about sets is a theorem of ZFC. From this it follows that ZFC is consistent if and only if NBG is consistent.

Limitations of Axiomatic Set Theory

The fact that NBG avoids the classical paradoxes and that there is no apparent way to derive any one of them in ZFC does not settle the question of the consistency of either theory. One method for establishing the consistency of an axiomatic theory is to give a model—i.e., an interpretation of the undefined terms in another theory such that the axioms become theorems of the other theory. If this other theory is consistent, then that under investigation must be consistent. Such consistency proofs are thus relative: the theory for which a model is given is consistent if that from which the model is taken is consistent. The method of models, however, offers no hope for proving the consistency of an axiomatic theory of sets. In the case of set theory and, indeed, of axiomatic theories generally, the alternative is a direct approach to the problem.

If T is the theory of which the (absolute) consistency is under investigation, this alternative means that the proposition "There is no sentence of T such that both it and its negation are theorems of T" must be proved. The mathematical theory (developed by the formalists) to cope with proofs about an axiomatic theory T is called proof theory, or metamathematics. It is premised upon the formulation of T as a formal axiomatic theory—i.e., the theory of inference (as well as T) must be axiomatized. It is then possible to present T in a purely symbolic form—i.e., as a formal language based on an alphabet the symbols of which are those for the undefined terms of T and those for the logical operators and connectives. A sentence in this language is a formula composed from the alphabet according to prescribed rules. The hope for metamathematics was that, by using only intuitively convincing, weak number-theoretic arguments (called finitary methods), and unimpeachable proofs of the consistency of such theories as axiomatic set theory could be given.

That hope suffered a severe blow in 1931 from a theorem proved by Kurt Gödel about any formal theory S that includes the usual vocabulary of elementary arithmetic. By coding the formulas of such a theory with natural numbers (now called Gödel numbers) and by talking about these numbers, Gödel was able to make the metamathematics of S become part of the arithmetic of S and hence expressible in S. The theorem in question asserts that the formula of S that expresses (via a coding) "S is consistent" in S is unprovable in S if S is consistent. Thus, if S is consistent, then the consistency of S cannot be proved within S; rather, methods beyond those that can be expressed or reflected in S must be employed. Because, in both ZFC and NBG, elementary arithmetic can be developed, Gödel's theorem applies to these two theories. Although there remains the theoretical possibility of a finitary proof of consistency that cannot be reflected in the foregoing systems of set theory, no hopeful, positive results have been obtained.

Other theorems of Gödel when applied to ZFC (and there are corresponding results for NBG) assert that, if the system is consistent, then (1) it contains a sentence such that neither it nor its negation is provable (such a sentence is called undecidable), (2) there is no algorithm (or iterative process) for deciding whether a sentence of ZFC is a theorem, and (3) these same statements hold for any consistent theory resulting from ZFC by the adjunction of further axioms or axiom schemas. Apparently ZFC can serve as a foundation for all of present-day mathematics because every mathematical theorem can be translated into and proved within ZFC or within extensions obtained by adding suitable axioms. Thus, the existence of undecidable sentences in each such theory points out an inevitable gap between the sentences that are true in mathematics and sentences those are provable within a single axiomatic theory. The fact that there is more to conceivable mathematics than can be captured by the axiomatic approach prompted the American logician Emil Post to comment in 1944 that "mathematical thinking is, and must remain, essentially creative."

Types of Sets

Empty Set or Null Set

A set which does not contain any element is called an empty set, or the null set or the void set and it is denoted by ∅ and is read as phi. In roster form, ∅ is denoted by {}. An empty set is a finite set, since the number of elements in an empty set is finite, i.e., 0.

For example:

- The set of whole numbers less than 0.
- Clearly there is no whole number less than 0.

Therefore, it is an empty set.

- $N = \{x : x \in N, 3 < x < 4\}$.
- Let $A = \{x : 2 < x < 3$, x is a natural number$\}$: Here A is an empty set because there is no natural number between 2 and 3.
- Let B = {x : x is a composite number less than 4}: Here B is an empty set because there is no composite number less than 4.

$\emptyset \neq \{0\}$ ∴ has no element.

{0} is a set which has one element 0.

The cardinal number of an empty set, i.e., $n(\emptyset) = 0$.

Singleton Set

A set which contains only one element is called a singleton set.

For example:

- A = {x : x is neither prime nor composite}: It is a singleton set containing one element, i.e., 1.
- B = {x : x is a whole number, x < 1}: This set contains only one element 0 and is a singleton set.
- Let A = {x : x ∈ N and x^2 = 4}: Here A is a singleton set because there is only one element 2 whose square is 4.
- Let B = {x : x is a even prime number}: Here B is a singleton set because there is only one prime number which is even, i.e., 2.

Finite Set

A set which contains a definite number of elements is called a finite set. Empty set is also called a finite set.

For example:

- The set of all colors in the rainbow.
- N = {x : x ∈ N, x < 7}.
- P = {2, 3, 5, 7, 11, 13, 17, ... 97}.

Infinite Set

The set whose elements cannot be listed, i.e., set containing never-ending elements is called an infinite set.

For example:

- Set of all points in a plane .
- A = {x : x ∈ N, x > 1}.
- Set of all prime numbers.
- B = {x : x ∈ W, x = 2n}.

All infinite sets cannot be expressed in roster form.

For example: The set of real numbers since the elements of this set do not follow any particular pattern.

Cardinal Number of a Set

The number of distinct elements in a given set A is called the cardinal number of A. It is denoted by n(A).

For example:

- A {x : x ∈ N, x < 5}.

  A = {1, 2, 3, 4}.

  Therefore, n(A) = 4.

- B = set of letters in the word ALGEBRA.

  B = {A, L, G, E, B, R}.

  Therefore, n(B) = 6.

Equivalent Sets

Two sets A and B are said to be equivalent if their cardinal number is same, i.e., n(A) = n(B). The symbol for denoting an equivalent set is '↔'.

For example:

- A = {1, 2, 3} Here n(A) = 3.

- B = {p, q, r} Here n(B) = 3.

  Therefore, A ↔ B.

Equal Sets

Two sets A and B are said to be equal if they contain the same elements. Every element of A is an element of B and every element of B is an element of A.

For example:

- A = {p, q, r, s}.

- B = {p, s, r, q}.

  Therefore, A = B.

Countable and Uncountable Sets

If AA is a finite set, there is a bijection F : n → A between a natural number n and A. Any such bijection gives a counting of the elements of A, namely, F(0) is the first element of

A, F(1) is the second, and so on. Thus, all finite sets are countable. An infinite set A is called countable if there is a bijection $F : \omega \to A$ between the set of natural numbers and A. The set NN of natural numbers is (trivially) countable. If AA is an infinite subset of ω, then AA is also countable: for let $F : \omega \to A$ be such that F(n) is the least element of AA that is not in the set $\{F(m) \in A : m < n\}$. Then F is a bijection.

Every infinite subset of a countable set is also countable: for suppose $F : \omega \to A$ is a bijection and $B \subseteq A$ is infinite. Then the set $\{n \in \omega : F(n) \in B\}$ is an infinite subset of ω, hence countable, and so there is a bijection $G : \omega \to \{n \in \omega : F(n) \in B\}$. Then the composition function $F \circ G : \omega \to B$ is a bijection.

The union of a countable set and a finite set is also countable. For given sets A and B, which without loss of generality we may assume they are disjoint, and given bijections $F : \omega \to A$ and $G : n \to B$, for some $n < \omega$, let $H : \omega \to A \cup B$ be the bijection given by: $H(m) = G(m)$, for every $m < n$, and $H(m) = F(m - n)$, for every $n \leq m$.

Suppose A and B are countable sets and $F : \omega \to A$ and $G : \omega \to B$ are bijections, then the function $H : \omega \to A \cup B$ consisting of all pairs $(2n, F(n))$, plus all pairs $(2n + 1, G(n))$ is a bijection.

Thus, the set $\mathbb{Z}$, being the union of two countable sets, namely,

$$\mathbb{N} \cup \{-1, -2, -3, -4, \ldots\}$$

is also countable.

The Cartesian product of two infinite countable sets is also countable. For suppose $F : \omega \to A$ and $G : \omega \to B$ are bijections. Then, using the fact that the function $J : \omega \times \omega \to \omega$ given by $J((m, n)) = 2m(2n + 1) - 1$ is a bijection, one has that the function $H : \omega \to A \times B$ given by $H(2m(2n + 1) - 1) = (F(m), G(n))$ is also a bijection.

Since any rational number is given by a pair of integers, i.e., a quotient $\frac{m}{n}$, where m, n $\in$ Z and $n \neq 0$, the set QQ of rational numbers is also countable.

However, Georg Cantor discovered that the set $\mathbb{R}$ of real numbers is not countable. For suppose, aiming for a contradiction, that $F : \omega \to R$ is a bijection. Let $a_0 = F(0)$. Choose the least k such that $a_0 < F(k)$ and put $b_0 = F(k)$. Given a_n and b_n, choose the least l such that $a_n < F(l) < b_n$, and put $a_{n+1} = F(l)$. And choose the least mm such that $a_{n+1} < F(m) < b_n$, and put $b_{n+1} = F(m)$. Thus, we have $a_0 < a_1 < a_2 < \cdots < b_2 < b_1 < b_0$. Now let a be the limit of the a_n. Then a is a real number different from F(n), all n, which is impossible because F is a bijection.

The existence of uncountable sets follows from a much more general fact, also discovered by Cantor. Namely, given any set A, the set of all its subsets, called the power set of A, and denoted by P(A), is not bijectable with A: for suppose that $F : A \to P(A)$ is a bijection. Then the subset $\{a \in A : a \notin F(a)\}$ of A is the value F(a) of some $a \in A$.

But then $a \in F(a)$ if and only if $a \notin F(a)$. Hence, if A is any infinite set, then P(A) is uncountable.

There are also uncountable ordinals. The set of all finite and countable ordinals is also an ordinal, called ω_1, and is the first uncountable ordinal. Similarly, the set of all ordinals that are bijectable with some ordinal less than or equal to ω_1 is also an ordinal, called ω_2, and is not bijectable with ω_1, and so on.

Cardinals

The cardinality, or size, of a finite set A is the unique natural number n such that there is a bijection $F : n \rightarrow A$.

In the case of infinite sets, their cardinality is given, not by a natural number, but by an infinite ordinal. However, in contrast with the finite sets, an infinite set A is bijectable with many different ordinal numbers. For example, the set $\mathbb{N}$ is bijectable with ω, but also with it successor $\omega \cup \{\omega\}$: by assigning 0 to ω and $n + 1$ to n, for all $n \in \omega$, we obtain a bijection between $\omega \cup \{\omega\}$ and ω. But since the ordinals are well-ordered, we may define the cardinality of an infinite set as the least ordinal that is bijectable with it.

In particular, the cardinality of an ordinal number α is the least ordinal κ that is bijectable with it. Notice that κ is not bijectable with any smaller ordinal, for otherwise so would be α. The ordinal numbers that are not bijectable with any smaller ordinal are called cardinal numbers. Thus, all natural numbers are cardinals, and so are ω, ω_1, ω_2, and so on. In general, given any cardinal κ, the set of all ordinals that are bijectable with some ordinal $\leq \kappa$ is also a cardinal; it is the smallest cardinal bigger than κ.

The infinite cardinals are represented by the letter aleph ($\aleph$) of the Hebrew alphabet. Thus, the smallest infinite cardinal is $\omega = \aleph_0$, the next one is $\omega_1 = \aleph_1$, which is the first uncountable cardinal, then comes $\omega_2 = \aleph_2$, etc.

The cardinality of any set A, denoted by $|A|$, is the unique cardinal number that is bijectable with A. We saw already that $|R|$ is uncountable, hence greater than $\aleph_0$, but it is not known what cardinal number it is. The conjecture that $|R|=\aleph_1$, formulated by Cantor in 1878, is the famous Continuum Hypothesis.

Combinatorial Set Theory

In mathematics, infinitary combinatorics, or combinatorial set theory, is an extension of ideas in combinatorics to infinite sets. Some of the things studied include continuous graphs and trees, extensions of Ramsey's theorem, and Martin's axiom. Recent developments concern combinatorics of the continuum and combinatorics on successors of singular cardinals.

Ramsey Theory for Infinite Sets

Write κ, λ for ordinals, m for a cardinal number and n for a natural number. Erdős and Rado introduced the notation:

$$\kappa \to (\lambda)^n_m$$

As a shorthand way of saying that every partition of the set $[\kappa]^n$ of n-element subsets of κ into m pieces has a homogeneous set of order type λ. A homogeneous set is in this case a subset of κ such that every n-element subset is in the same element of the partition. When m is 2 it is often omitted.

Assuming the axiom of choice, there are no ordinals κ with $\kappa \to (\omega)^\omega$, so n is usually taken to be finite. An extension where n is almost allowed to be infinite is the notation:

$$\kappa \to (\lambda)^{<\omega}_m$$

Which is a shorthand way of saying that every partition of the set of finite subsets of κ into m pieces has a subset of order type λ such that for any finite n, all subsets of size n are in the same element of the partition. When m is 2 it is often omitted.

Another variation is the notation:

$$\kappa \to (\lambda, \mu)^n$$

Which is a shorthand way of saying that every coloring of the set $[\kappa]^n$ of n-element subsets of κ with 2 colors has a subset of order type λ such that all elements of $[\lambda]^n$ have the first color, or a subset of order type μ such that all elements of $[\mu]^n$ have the second color.

Some properties of this include: (in what follows κ is a cardinal).

- $\aleph_0 \to (\aleph_0)^n_k$ for all finite n and k (Ramsey's theorem)
- $\beth_n^+ \to (\aleph_1)^{n+1}_{\aleph_0}$ (Erdős–Rado theorem)
- $2^\kappa \not\to (\kappa^+)^2$ (Sierpiński theorem)
- $2^\kappa \not\to (3)^2_\kappa$
- $\kappa \to (\kappa, \aleph_0)^2$ (Erdős–Dushnik–Miller theorem)

In choiceless universes, partition properties with infinite exponents may hold, and some of them are obtained as consequences of the axiom of determinacy (AD). For example, Donald A. Martin proved that AD implies:

$$\aleph_1 \to (\aleph_1)^{\aleph_1}_2$$

Descriptive Set Theory

In mathematical logic, descriptive set theory (DST) is the study of certain classes of "well-behaved" subsets of the real line and other Polish spaces. As well as being one of the primary areas of research in set theory, it has applications to other areas of mathematics such as functional analysis, ergodic theory, the study of operator algebras and group actions, and mathematical logic.

Polish Spaces

Descriptive set theory begins with the study of Polish spaces and their Borel sets.

A Polish space is a second-countable topological space that is metrizable with a complete metric. Heuristically, it is a complete separable metric space whose metric has been "forgotten". Examples include the real line $\mathbb{R}$, the Baire space $\mathcal{N}$, the Cantor space $\mathcal{C}$, and the Hilbert cube $I^{\mathbb{N}}$.

Universality Properties

The class of Polish spaces has several universality properties, which show that there is no loss of generality in considering Polish spaces of certain restricted forms.

- Every Polish space is homeomorphic to a G_δ subspace of the Hilbert cube, and every G_δ subspace of the Hilbert cube is Polish.
- Every Polish space is obtained as a continuous image of Baire space; in fact every Polish space is the image of a continuous bijection defined on a closed subset of Baire space. Similarly, every compact Polish space is a continuous image of Cantor space.

Because of these universality properties, and because the Baire space $\mathcal{N}$ has the convenient property that it is homeomorphic to $\mathcal{N}^{\omega}$, many results in descriptive set theory are proved in the context of Baire space alone.

Borel Sets

The class of Borel sets of a topological space X consists of all sets in the smallest σ-algebra containing the open sets of X. This means that the Borel sets of X are the smallest collection of sets such that:

- Every open subset of X is a Borel set.
- If A is a Borel set, so is $X \setminus A$. That is, the class of Borel sets are closed under complementation.

- If An is a Borel set for each natural number n, then the union $\bigcup A_n$ is a Borel set. That is, the Borel sets are closed under countable unions.

A fundamental result shows that any two uncountable Polish spaces X and Y are Borel isomorphic: there is a bijection from X to Y such that the preimage of any Borel set is Borel, and the image of any Borel set is Borel. This gives additional justification to the practice of restricting attention to Baire space and Cantor space, since these and any other Polish spaces are all isomorphic at the level of Borel sets.

Borel Hierarchy

Each Borel set of a Polish space is classified in the Borel hierarchy based on how many times the operations of countable union and complementation must be used to obtain the set, beginning from open sets. The classification is in terms of countable ordinal numbers. For each nonzero countable ordinal α there are classes Σ^0_α, Π^0_α and Δ^0_α.

- Every open set is declared to be Σ^0_1.
- A set is declared to be Π^0_α if and only if its complement is Π^0_α.
- A set A is declared to be Σ^0_δ, $\delta > 1$, if there is a sequence $\langle Ai \rangle$ of sets, each of which is $\Pi^0_{\lambda(i)}$ for some $\lambda(i) < \delta$, such that $A = \bigcup A_i$.
- A set is Δ^0_α if and only if it is both Σ^0_α and Π^0_α.

A theorem shows that any set that is Σ^0_α or Π^0_α is $\Delta^0_{\alpha+1}$, and any Δ^0_β set is both Σ^0_α and Π^0_α for all $\alpha > \beta$. Thus the hierarchy has the following structure, where arrows indicate inclusion.

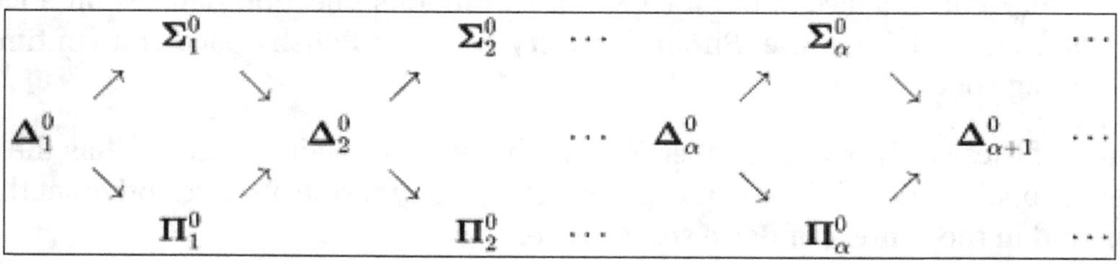

Regularity Properties of Borel Sets

Classical descriptive set theory includes the study of regularity properties of Borel sets. For example, all Borel sets of a Polish space have the property of Baire and the perfect set property. Modern descriptive set theory includes the study of the ways in which these results generalize, or fail to generalize, to other classes of subsets of Polish spaces.

Analytic and Coanalytic Sets

Just beyond the Borel sets in complexity are the analytic sets and coanalytic sets.

A subset of a Polish space X is analytic if it is the continuous image of a Borel subset of some other Polish space. Although any continuous preimage of a Borel set is Borel, not all analytic sets are Borel sets. A set is coanalytic if its complement is analytic.

Projective Sets and Wadge Degrees

Many questions in descriptive set theory ultimately depend upon set-theoretic considerations and the properties of ordinal and cardinal numbers. This phenomenon is particularly apparent in the projective sets. These are defined via the projective hierarchy on a Polish space X:

- A set is declared to be Σ_1^1 if it is analytic.
- A set is Π_1^1 if it is coanalytic.
- A set A is Σ_{n+1}^1 if there is a Π_n^1 subset B of $X \times X$ such that A is the projection of B to the first coordinate.
- A set A is Π_{n+1}^1 if there is a Σ_n^1 subset B of $X \times X$ such that A is the projection of B to the first coordinate.
- A set is Π_n^1 if it is both Π_n^1 and Σ_n^1.

As with the Borel hierarchy, for each n, any Δ_n^1 set is both Σ_{n+1}^1 and Π_{n+1}^1.

The properties of the projective sets are not completely determined by ZFC. Under the assumption V = L, not all projective sets have the perfect set property or the property of Baire. However, under the assumption of projective determinacy, all projective sets have both the perfect set property and the property of Baire. This is related to the fact that ZFC proves Borel determinacy, but not projective determinacy.

More generally, the entire collection of sets of elements of a Polish space X can be grouped into equivalence classes, known as Wadge degrees that generalize the projective hierarchy. These degrees are ordered in the Wadge hierarchy. The axiom of determinacy implies that the Wadge hierarchy on any Polish space is well-founded and of length Θ, with structure extending the projective hierarchy.

Borel Equivalence Relations

A contemporary area of research in descriptive set theory studies Borel equivalence relations. A Borel equivalence relation on a Polish space X is a Borel subset of {\displaystyle X\times X}X\times X that is an equivalence relation on X.

Effective Descriptive Set Theory

The area of effective descriptive set theory combines the methods of descriptive set theory with those of generalized recursion theory (especially hyperarithmetical theory). In particular, it focuses on lightface analogues of hierarchies of classical descriptive set theory. Thus the hyperarithmetic hierarchy is studied instead of the Borel hierarchy, and the analytical hierarchy instead of the projective hierarchy. This research is related to weaker versions of set theory such as Kripke–Platek set theory and second-order arithmetic.

Lightface		Boldface	
$\Sigma^0_0 = \Pi^0_0 = \Delta^0_0$ (sometimes the same as Δ^0_1)		$\Sigma^0_0 = \Pi^0_0 = \Delta^0_0$ (if defined)	
Δ^0_1 = recursive		Δ^0_1 = clopen	
Σ^0_2 = recursively enumerable	Π^0_1 = o-recursively enumerable	Σ^0_2 = G = open	Π^0_1 = F = closed
Δ^0_2		Δ^0_2	
Σ^0_2	Π^0_2	$\Sigma^0_2 = F_\sigma$	$\Pi^0_2 = G_\delta$
Δ^0_3		Δ^0_3	
Σ^0_3	Π^0_3	$\Sigma^0_3 = G_{\delta\sigma}$	$\Pi^0_3 = F_{\sigma\delta}$
⋮		⋮	
$\Sigma^0_{<\omega} = \Pi^0_{<\omega} = \Delta^0_{<\omega} = \Sigma^1_0 = \Pi^1_0 = \Delta^1_0 =$ arithmetical		$\Sigma^0_{<\omega} = \Pi^0_{<\omega} = \Delta^0_{<\omega} = \Sigma^1_0 = \Pi^1_0 = \Delta^1_0 =$ Boldface arithmetical	
⋮		⋮	
Δ^0_α (α recursive)		Δ^0_α (α countable)	
Σ^0_α	Π^0_α	Σ^0_α	Π^0_α
⋮		⋮	
$\Sigma^0_{\omega_1^{CK}} = \Pi^0_{\omega_1^{CK}} = \Delta^0_{\omega_1^{CK}} = \Delta^1_1 =$ hyperarithmetical		$\Sigma^0_{\omega_1} = \Pi^0_{\omega_1} = \Delta^0_{\omega_1} = \Delta^1_1 =$ Borel	
$\Sigma^1_1 =$ lightface analytic	$\Pi^1_1 =$ lightface coanalytic	$\Sigma^1_1 =$ A = analytic	$\Pi^1_1 =$ CA = coanalytic
Δ^1_2		Δ^1_2	
Σ^1_2	Π^1_2	Σ^1_2 = PCA	Π^1_2 = CPCA

Δ^1_3		Δ^1_3	
Σ^1_3	Π^1_3	$\Sigma^1_3 = \text{PCPCA}$	$\Pi^1_3 = \text{CPCPCA}$
⋮		⋮	
$\Sigma^1_{<\omega} = \Pi^1_{<\omega} = \Delta^1_{<\omega} = \Sigma^2_0 = \Pi^2_0 = \Delta^2_0 =$ analytical		$\Sigma^1_{<\omega} = \Pi^1_{<\omega} = \Delta^1_{<\omega} = \Sigma^2_0 = \Pi^2_0 = \Delta^2_0 = \text{P} =$ projective	
⋮		⋮	

Fuzzy Set Theory

In mathematics, fuzzy sets (aka uncertain sets) are somewhat like sets whose elements have degrees of membership. Fuzzy sets were introduced independently by Lotfi A. Zadeh and Dieter Klaua [de] in 1965 as an extension of the classical notion of set. At the same time, Salii defined a more general kind of structure called an L-relation, which he studied in an abstract algebraic context. Fuzzy relations, which are used now in different areas, such as linguistics, decision-making, and clustering, are special cases of L-relations when L is the unit interval [0, 1].

In classical set theory, the membership of elements in a set is assessed in binary terms according to a bivalent condition — an element either belongs or does not belong to the set. By contrast, fuzzy set theory permits the gradual assessment of the membership of elements in a set; this is described with the aid of a membership function valued in the real unit interval [0, 1]. Fuzzy sets generalize classical sets, since the indicator functions (aka characteristic functions) of classical sets are special cases of the membership functions of fuzzy sets, if the latter only take values 0 or 1. In fuzzy set theory, classical bivalent sets are usually called crisp sets. The fuzzy set theory can be used in a wide range of domains in which information is incomplete or imprecise, such as bioinformatics.

A fuzzy set is a pair (U, m) where U is a set and $m : U \to [0,1]$ a membership function. The reference set U (sometimes denoted by Ω or X) is called universe of discourse, and for each $x \in U$, the value m(x) is called the grade of membership of x in (U, m). The function $m = \mu_A$ is called the membership function of the fuzzy set A = (U, m).

For a finite set (U, m), the fuzzy set (U, m) is often denoted by (x_{1})/x_{1},\dots ,m(x_{n})/x_{n}\}.

Let, $x \in U$. Then x is called:

- Not included in the fuzzy set (U, m) if m(x) = 0 (no member).

- Fully included if m(x) = 1 (full member).
- Partially included if 0 < m(x) < 1 (fuzzy member).

The (crisp) set of all fuzzy sets on a universe U is denoted with SF(U) (or sometimes just F(U).

Crisp Sets Related to a Fuzzy Set

For any fuzzy set A = (U, m) and $\alpha \in [0, 1]$ the following crisp sets are defined:

- $A^{\geq\alpha} = A_\alpha = \{x \in U \mid m(x) \geq \alpha\}$ is called its α-cut (aka α-level set).
- $A^{>\alpha} = A'_\alpha = \{x \in U \mid m(x) > \alpha\}$ is called its strong α-cut (aka strong α-level set).
- $S(A) = \text{Supp}(A) = A^{>0} = \{x \in U \mid m(x) > 0\}$ is called its support.
- $C(A) = \text{Core}(A) = A^{=1} = \{x \in U \mid m(x) = 1\}$ is called its core (or sometimes kernel $\text{Kern}(A)$).

Other Definitions

- A fuzzy set $A = (U, m)$ is empty $A = \varnothing$ iff (if and only if):

  $$x \in U : \mu_A(x) = m(x) = 0$$

- Two fuzzy sets A and B are equal (A = B) iff:

  $$\forall x \in U : \mu_A(x) = \mu_B(x)$$

- A fuzzy set A is included in a fuzzy set $B(A \subseteq B)$ iff:

  $$\forall x \in U : \mu_A(x) \leq \mu_B(x)$$

- For any fuzzy set A, any element $x \in U$ in U that satisfies:

  $$\mu_A(x) = 0.5$$

is called a crossover point.

- Given a fuzzy set A, any $\alpha \in [0,1]$, for which $A^{=\alpha} = \{x \in U \mid \mu_A(x) = \alpha\}$ is not empty, is called a level of A.
- The level set of A is the set of all levels $\alpha \in [0,1]$ representing distinct cuts. It is the target set (aka range or image) of μ_A:

  $$\Lambda_A = \{\alpha \in [0,1] \mid A^{=\alpha} \neq \varnothing\} = \{\alpha \in [0,1] \mid \exists$$
  $$x \in U : \mu_A(x) = \alpha\} = \mu_A(U)$$

- For a fuzzy set A, its height is given by:

$$\mathrm{Hgt}(A) = \sup\{\mu_A(x) \mid x \in U\} = \sup(\mu_A(U))$$

Where sup denotes the supremum, which is known to exist because 1 is an upper bound. If U is finite, we can simply replace the supremum by the maximum.

- A fuzzy set A is said to be normalized iff:

$$\mathrm{Hgt}(A) = 1$$

In the finite case, where the supremum is a maximum, this means that at least one element of the fuzzy set has full membership. A non-empty fuzzy set A may be normalized with result $\tilde{A}$ by dividing the membership function of the fuzzy set by its height:

$$\forall x \in U : \mu_{\tilde{A}}(x) = \mu_A(x) / \mathrm{Hgt}(A)$$

Besides similarities this differs from the usual normalization in that the normalizing constant is not a sum.

- For fuzzy sets A of real numbers ($U \subseteq R$) having a support with an upper and a lower bound, the width is defined as:

$$\mathrm{Width}(A) = \sup(\mathrm{Supp}(A)) - \inf(\mathrm{Supp}(A))$$

This does always exist for a bounded reference set U, including when U is finite.

In case that Supp(A) is a finite or closed set, the width is just:

$$\mathrm{Width}(A) = \max(\mathrm{Supp}(A)) - \min(\mathrm{Supp}(A))$$

In the n-dimensional case ($U \subseteq R^n$) the above can be replaced by the n-dimensional volume of Supp(A).

In general, this can be defined given any measure on U, for instance by integration (e. g. Lebesgue integration) of Supp(A).

- A real fuzzy set {\displaystyle A}A ($U \subseteq R$) is said to be convex (in the fuzzy sense, not to be confused with a crisp convex set), iff:

$$\forall x, y \in U, \forall \lambda \in [0,1] : \mu_A(\lambda x + (1-\lambda)y) \geq \min(\mu_A(x), \mu_A(y))$$

Without loss of generality, we may take $x \leq y$, which gives the equivalent formulation:

$$\forall z \in [x, y] : \mu_A(z) \geq \min(\mu_A(x), \mu_A(y))$$

This definition can be extended to one for a general topological space U: we say the fuzzy set A is convex when, for any subset Z of U, the condition:

$$\forall z \in Z : \mu_A(z) \geq \inf(\mu_A(\partial Z))$$

holds, where ∂Z denotes the boundary of Z and $f(X) = \{f(x) \mid x \in X\}$ denotes the image of a set X (here ∂Z) under a function f (here μ_A).

Disjoint Fuzzy Sets

In contrast to the general ambiguity of intersection and union operations, there is clearness for disjoint fuzzy sets: Two fuzzy sets A, B are disjoint iff:

$$\forall x \in U : \mu_A(x) = 0 \vee \mu_B(x) = 0$$

Which is equivalent to:

$$x \in U : \mu_A(x) > 0 \wedge \mu_B(x) > 0$$

And also equivalent to:

$$\forall x \in U : \min(\mu_A(x), \mu_B(x)) = 0$$

We keep in mind that min/max is a t/s-norm pair, and any other will do the job here as well.

Fuzzy sets are disjoint, iff their supports are disjoint according to the standard definition for crisp sets.

For disjoint fuzzy sets A, B any intersection will give ∅, and any union will give the same result, which is denoted as:

$$A \dot{\cup} B = A \cup B$$

With its membership function given by:

$$\forall x \in U : \mu_{A \dot{\cup} B}(x) = \mu_A(x) + \mu_B(x)$$

Note that only one of both summands is greater than zero.

For disjoint fuzzy sets A, B the following holds true:

$$\mathrm{Supp}(A \dot{\cup} B) = \mathrm{Supp}(A) \cup \mathrm{Supp}(B)$$

This can be generalized to finite families of fuzzy sets as follows: Given a family $A = (A_i)_{i \in I}$ of fuzzy sets with Index set I (e.g. I = {1,2,3,..., n}). This family is (pairwise) disjointing iff:

$$\forall x \in U \exists \text{ at most one } i \in I : \mu_{A_i}(x) > 0$$

A family of fuzzy sets $A = (A_i)_{i \in I}$ is disjoint; iff the family of underlying supports $\text{Supp}^\circ A = (\text{Supp}(A_i))_{i \in I}$ is disjoint in the standard sense for families of crisp sets.

Independent of the t/s-norm pair, intersection of a disjoint family of fuzzy sets will give ∅ again, while the union has no ambiguity:

$$\dot{\bigcup_{i \in I}} A_i = \bigcup_{i \in I} A_i$$

With its membership function given by:

$$\forall x \in U: \mu_{\quad_{i \in I}} (x) = \sum_{i \in I} \mu \ (x)$$

Again only one of the summands is greater than zero.

For disjoint families of fuzzy sets $A = (A_i)_{i \in I}$ the following holds true:

$$\text{Supp}\left(\dot{\bigcup_{i \in I}} A_i \right) = \bigcup_{i \in I} \text{Supp}(A_i)$$

Scalar Cardinality

For a fuzzy set A with finite Supp(A) (i. e. a 'finite fuzzy set'), its cardinality (aka scalar cardinality or sigma-count) is given by:

$$\text{Card}(A) = \text{sc}(A) = |A| = \sum_{x \in U} \mu_A(x)$$

In case that U itself is a finite set, the relative cardinality is given by:

$$\text{RelCard}(A) = ||A|| = \text{sc}(A) / |U| = |A| / |U|$$

This can be generalized for the divisor to be an non-empty fuzzy set: For fuzzy sets A, G with G ≠ ∅, we can define the relative cardinality by:

$$\text{RelCard}(A, G) = \text{sc}(A | G) = \text{sc}(A \cap G) / \text{sc}(G)$$

which looks very similar to the expression for conditional probability.

- $\text{sc}(G) > 0$ here.
- The result may depend on the specific intersection (t-norm) chosen.
- For G = U the result is unambiguous and resembles the prior definition.

Distance and Similarity

For any fuzzy set A the membership function $\mu_A : U \rightarrow U$ can be regarded as a family $\mu_A = (\mu_A(x))_{x \in U} \in [0,1]^U$. The latter is a metric space with several metrics d known. A metric can be derived from a norm (vector norm) $\|\|$ via:

$$d(\alpha,\beta) = \|\alpha - \beta\|$$

For instance, if U is finite, i. e. $U = \{x_1, x_2, ... x_n\}$, such a metric may be defined by:

$d(\alpha,\beta) := \max\{|\alpha(x_i) - \beta(x_i)| \, | \, i = 1..n\}$ where α and β are sequences of real numbes between 0 and 1.

For infinite U, the maximum can be replaced by a supremum. Because fuzzy sets are unambiguously defined by their membership function, this metric can be used to measure distances between fuzzy sets on the same universe:

$$d(A,B) := d(\mu_A, \mu_B)$$

which becomes in the above sample:

$$d(A,B) = \max\{|\mu_A(x_i) - \mu_B(x_i)| \, | \, i = 1...n\}$$

Again for infinite U the maximum must be replaced by a supremum. Other distances (like the canonical 2-norm) may diverge, if infinite fuzzy sets are too different, e .g ∅ and U.

Similarity measures (here denoted by S) may then be derived from the distance, e. g. after a proposal by Koczy:

$S = 1/(1 + d(A,B))$ if d(A,B) is finite, 0 else,

Or after Williams an Steele:

$S = \exp(-\alpha d(A,B))$ if $d(A,B)$ is finite, 0 else,

Where α >0 is a steepness parameter and exp(x) = e^x.

Another definition for interval valued (rather 'fuzzy') similarity measures ζ is provided by Beg and Ashraf as well.

L-fuzzy Sets

Sometimes, more general variants of the notion of fuzzy set are used, with membership functions taking values in a (fixed or variable) algebra or structure L of a given kind; usually it is required that L be at least a poset or lattice. These are usually called L-fuzzy sets, to distinguish them from those valued over the unit interval. The usual membership functions with values in [0, 1] are then called [0, 1]-valued membership functions. These kinds of generalizations were first considered in 1967 by Joseph Goguen, who was a student of Zadeh. A classical corollary may be indicating truth and membership values by {f, t} instead of {0,1}.

An extension of fuzzy sets has been provided by Atanassov and Baruah. An intuitionistic fuzzy set (IFS) A is characterized by two functions:

- $\mu_A(x)$ - Degree of membership of x.
- $\nu_A(x)$ - Degree of non-membership of x.

With functions $\mu_A, \nu_A : U \mapsto [0,1]$ with $\forall x \in U : \mu_A(x) + \nu_A(x) \leq 1$.

This resembles a situation like some person denoted by x voting:

- For a proposal A ($\mu_A(x) = 1, \nu_A(x) = 0$).
- Against it ($\mu_A(x) = 0, \nu_A(x) = 1$).
- Abstain from voting ($\mu_A(x) = \nu_A(x) = 0$).

After all, we have a percentage of approvals, a percentage of denials, and a percentage of abstentions.

For this situation, special 'intuitive fuzzy' negators, t- and s-norms can be provided. With $D^* = \{(\alpha,\beta) \in [0,1]^2 \mid \alpha + \beta = 1\}$ and by combining both functions to $(\mu_A, \nu_A) : U \to D^*$ this situation resembles a special kind of L-fuzzy sets.

Once more, this has been expanded by defining picture fuzzy sets (PFS) as follows: A PFS A is characterized by three functions mapping U to [0, 1]: μ_A, η_A, ν_A, 'degree of positive membership', 'degree of neutral membership', and 'degree of negative membership' respectively and additional condition $\forall x \in U : \mu_A(x) + \eta_A(x) + \nu_A(x) \leq 1$. This expands the voting sample above by an additional possibility 'refusal of voting'.

With $D^* = \{(\alpha,\beta,\gamma) \in [0,1]^3 \mid \alpha + \beta + \gamma = 1\}$ and special 'picture fuzzy' negators, t- and s-norms this resembles just another type of L-fuzzy sets.

Neutrosophic Fuzzy Sets

The concept of IFS has been extended into two major models. The two extensions of IFS are neutrosophic fuzzy sets and Pythagorean fuzzy sets.

Neutrosophic fuzzy sets were introduced by Smarandache in 1998. Like IFS, neutrosophic fuzzy sets have the previous two functions: one for membership $\mu_A(x)$ and another for non-membership $\nu_A(x)$. The major difference is that neutrosophic fuzzy sets have one more function: for indeterminate $i_A(x)$. This value indicates that the degree of undecidedness that the entity x belongs to the set. This concept of having indeterminate $i_A(x)$ value can be particularly useful when one cannot be very confident on the membership or non-membership values for item x. In summary, neutrosophic fuzzy sets are associated with the following functions:

- $\mu_A(x)$: Degree of membership of x.
- $\nu_A(x)$: Degree of non-membership of x.
- $i_A(x)$: Degree of indeterminate value of x.

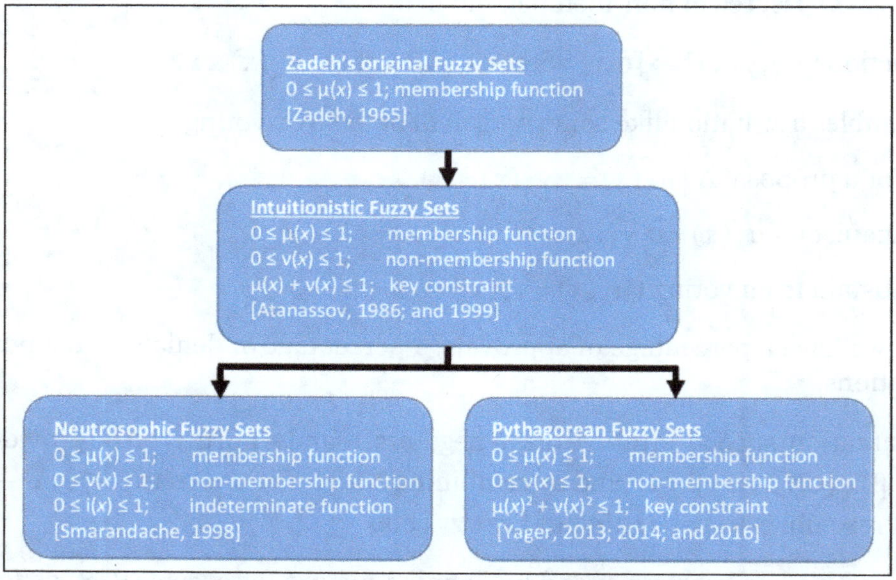

Some Key Developments in the Introduction of Fuzzy Set Concepts.

Pythagorean Fuzzy Sets

The other extension of IFS is what is known as Pythagorean fuzzy sets. Pythagorean fuzzy sets are more flexible than IFS. IFS are based on the constrain is $\mu_A(x)+\nu_A(x)\leq 1$, which can be considered as too restrictive in some occasions. This is why Yager proposed the concept of Pythagorean fuzzy sets. Such sets satisfy the constrain of $\mu_A(x)^2+\nu_A(x)^2\leq 1$, which is reminiscent of the Pythagorean theorem. Pythagorean fuzzy sets can be applicable to real life applications in which the previous condition of $\mu_A(x)+\nu_A(x)\leq 1$ is not valid. However, the less restrictive condition of $\mu_A(x)^2+\nu_A(x)^2\leq 1$ may be suitable in more domains.

Fuzzy Categories

The use of set membership as key components of category theory can be generalized to fuzzy sets. This approach which initiated in 1968 shortly after the introduction of fuzzy set theory led to the development of "Goguen categories" in the 21st century. In these categories, rather than using two valued set membership, more general intervals are used, and may be lattices as in L-fuzzy sets.

Fuzzy Relation Equation

The fuzzy relation equation is an equation of the form $A \cdot R = B$, where A and B are fuzzy sets, R is a fuzzy relation, and $A \cdot R$ stands for the composition of A with R.

Entropy

A measure d of fuzzyness for fuzzy sets of universe U should fulfill the following conditions for all $x \in U$:

- $d(A) = 0$, if A is a crisp set: $\mu_A(x) \in \{0,1\}$.
- $d(A) = 0$ has a unique maximum iff $\forall x \in U : \mu_A(x) = 0.5$.
- $d(A) \geq d(B)$ iff:
    - $\mu_a(x) \leq \mu_B(x)$ for $\mu_A(x) \leq 0.5$.
    - $\mu_a(x) \geq \mu_B(x)$ for $\mu_A(x) \geq 0.5$.

which means that B is 'crisper' than A.

$$d(\neg A) = d(A)$$

In this case d(A) is called the entropy of the fuzzy set A.

For finite $U = \{x_1, x_2, \ldots x_n\}$ the entropy of a fuzzy set A is given by:

$$d(A) = H(A) + H(\neg A)$$

$$H(A) = -k\sum_{i=1}^{n} \mu_A(x_i) \ln \mu_A(x_i)$$

Or just,

$$d(A) = -k\sum_{i=1}^{n} S(\mu_A(x_i))$$

where $S(x) = H_e(x)$ is Shannon's function (natural entropy function).

$$S(\alpha) = -\alpha \ln \alpha - (1-\alpha)\ln(1-\alpha), \alpha \in [0,1]$$

And k is a constant depending on the measure unit and the logarithm base (here: e) used. Physical interpretation of k is the Boltzmann constant k^B.

Let A be a fuzzy set with a continuous membership function (fuzzy variable). Then:

$$H(A) = -k\int_{-\infty}^{\infty} Cr\{A \geq t\} \ln Cr\{A \geq t\} dt$$

and its entropy is:

$$d(A) = -k\int_{-\infty}^{\infty} S(Cr\{A \geq t\}) dt$$

Fuzzy Set Operations

A fuzzy set operation is an operation on fuzzy sets. These operations are generalization of crisp set operations. There is more than one possible generalization. The most widely used operations are called standard fuzzy set operations. There are three operations: fuzzy complements, fuzzy intersections, and fuzzy unions.

Standard Fuzzy Set Operations

Let A and B be fuzzy sets that $A,B \subseteq U$, u is any element (e.g. value) in the U universe: $u \in U$.

- Standard complement: $\mu_{\neg A}(u) = 1 - \mu_A(u)$.

The complement is sometimes denoted by ∁A or A^C instead of $\neg A$.

- Standard intersection: $\mu_{A \cap B}(u) = \min\{\mu_A(u), \mu_B(u)\}$.
- Standard union: $\mu_{A \cup B}(u) = \max\{\mu_A(u), \mu_B(u)\}$.

In general, the triple (i,u,n) is called De Morgan Triplet iff:

- i is a t-norm.
- u is a t-conorm (aka s-norm).
- n is a strong negator.

So that for all $x, y \in [0, 1]$ the following holds true:

$$u(x,y) = n(i(n(x), n(y)))$$

(Generalized De Morgan relation).

Fuzzy Complements

$\mu_A(x)$ is defined as the degree to which x belongs to A. Let ∁A denote a fuzzy complement of A of type c. Then $\mu_{CA}(x)$ is the degree to which x belongs to CA, and the degree to which x does not belong to A. ($\mu_A(x)$ is therefore the degree to which x does not belong to CA.) Let a complement CA be defined by a function:

$$c:[0,1]\rightarrow[0,1]$$

For all $x\in U:\mu_{CA}(x) = c(\mu_A(x))$.

Axioms for Fuzzy Complements

- Boundary condition: $c(0)=1$ and $c(1)=0$.
- Monotonicity: For all a, b $\in$ [0, 1], if a < b, then c(a) > c(b).
- Continuity: c is continuous function.
- Involutions: c is an involution, which means that c(c(a)) = a for each a $\in$ [0,1]

c is a strong negator (aka fuzzy complement).

A function c satisfying axioms c1 and c2 has at least one fixpoint a* with c(a*) = a*, and if axiom c3 is fulfilled as well there is exactly one such fixpoint. For the standard negator c(x) = 1 - x the unique fixpoint is a* = 0.5.

Fuzzy Intersections

The intersection of two fuzzy sets A and B is specified in general by a binary operation on the unit interval, a function of the form:

$$i:[0,1]\times[0,1] \rightarrow [0,1]$$

For all x $\in$ U: $\mu_{A\cap B}(x) = i\,[\mu_A(x), \mu_B(x)]$.

Axioms for Fuzzy Intersection

- Boundary condition: i (a, 1) = a.
- Monotonicity: b $\leq$ d implies i (a, b) $\leq$ i (a, d).
- Commutativity: i (a, b) = i (b, a).
- Associativity: i (a, i (b, d)) = i (i (a, b), d).
- Continuity: i is a continuous function.
- Subidempotency: i (a, a) $\leq$ a.
- Strict monotonicity: $i(a_1, b_1) \leq i(a_2, b_2)$ if $a_1 \leq a_2$ and $b_1 \leq b_2$.

Axioms i1 up to i4 define a t-norm (aka fuzzy intersection). The standard t-norm min is the only idempotent t-norm (that is, i $(a_1, a_1) = a$ for all $a \in [0,1]$).

Fuzzy Unions

The union of two fuzzy sets A and B is specified in general by a binary operation on the unit interval function of the form:

$$u:[0,1]\times[0,1] \rightarrow [0,1].$$

For all $x \in U$: $\mu A \cup B(x) = u[\mu A(x), \mu B(x)]$.

Axioms for Fuzzy Union

- Boundary condition: $u(a, 0) = u(0, a) = a$.
- Monotonicity: $b \leq d$ implies $u(a, b) \leq u(a, d)$.
- Commutativity: $u(a, b) = u(b, a)$.
- Associativity: $u(a, u(b, d)) = u(u(a, b), d)$.
- Continuity: u is a continuous function.
- Superidempotency: $u(a, a) \geq a$.
- Strict monotonicity: $a_1 < a_2$ and $b_1 < b_2$ implies $u(a_1, b_1) < u(a_2, b_2)$.

Axioms u1 up to u4 define a t-conorm (aka s-norm or fuzzy intersection). The standard t-conorm max is the only idempotent t-conorm (i. e. u (a1, a1) = a for all $a \in [0,1]$).

Aggregation Operations

Aggregation operations on fuzzy sets are operations by which several fuzzy sets are combined in a desirable way to produce a single fuzzy set.

Aggregation operation on n fuzzy set ($2 \leq n$) is defined by a function:

$$h:[0,1]^n \rightarrow [0,1]$$

Axioms for Aggregation Operations Fuzzy Sets

- Boundary condition: $h(0, 0, ..., 0) = 0$ and $h(1, 1, ..., 1) =$ one.
- Monotonicity: For any pair $<a_1, a_2, ..., a_n>$ and $<b_1, b_2, ..., b_n>$ of n-tuples such that $a_i, b_i \in [0,1]$ for all $i \in N_n$, if $a_i \leq b_i$ for all $i \in N_n$, then $h(a_1, a_2, ..., a_n) \leq h(b_1, b_2, ..., b_n)$; that is, h is monotonic increasing in all its arguments.
- Continuity: h is a continuous function.

Fuzzy Number

A fuzzy number is a generalization of a regular, real number in the sense that it does not refer to one single value but rather to a connected set of possible values, where each possible value has its own weight between 0 and 1. This weight is called the membership function. A fuzzy number is thus a special case of a convex, normalized fuzzy set of the real line. Just like Fuzzy logic is an extension of Boolean logic (which uses absolute truth and falsehood only, and nothing in between), fuzzy numbers are an extension of real numbers. Calculations with fuzzy numbers allow the incorporation of uncertainty on parameters, properties, geometry, initial conditions, etc. The arithmetic calculations on fuzzy numbers are implemented using fuzzy arithmetic operations, which can be done by two different approaches: (1) interval arithmetic approach; and (2) the extension principle approach.

Fuzzy Logic

Fuzzy logic is a form of many-valued logic in which the truth values of variables may be any real number between 0 and 1 both inclusive. It is employed to handle the concept of partial truth, where the truth value may range between completely true and completely false. By contrast, in Boolean logic, the truth values of variables may only be the integer values 0 or 1.

The term fuzzy logic was introduced with the 1965 proposal of fuzzy set theory by Lotfi Zadeh. Fuzzy logic had however been studied since the 1920s, as infinite-valued logic—notably by Łukasiewicz and Tarski.

Fuzzy logic is based on the observation that people make decisions based on imprecise and non-numerical information. Fuzzy models or sets are mathematical means of representing vagueness and imprecise information (hence the term fuzzy). These models have the capability of recognising, representing, manipulating, interpreting, and utilising data and information that are vague and lack certainty.

Fuzzy logic has been applied to many fields, from control theory to artificial intelligence.

Classical logic only permits conclusions which are either true or false. However, there are also propositions with variable answers, such as one might find when asking a group of people to identify a color. In such instances, the truth appears as the result of reasoning from inexact or partial knowledge in which the sampled answers are mapped on a spectrum.

Both degrees of truth and probabilities range between 0 and 1 and hence may seem similar at first, but fuzzy logic uses degrees of truth as a mathematical model of vagueness, while probability is a mathematical model of ignorance.

Applying Truth Values

A basic application might characterize various sub-ranges of a continuous variable. For instance, a temperature measurement for anti-lock brakes might have several separate membership functions defining particular temperature ranges needed to control the brakes properly. Each function maps the same temperature value to a truth value in the 0 to 1 range. These truth values can then be used to determine how the brakes should be controlled.

Linguistic Variables

While variables in mathematics usually take numerical values, in fuzzy logic applications, non-numeric values are often used to facilitate the expression of rules and facts.

A linguistic variable such as age may accept values such as young and its antonym old. Because natural languages do not always contain enough value terms to express a fuzzy value scale, it is common practice to modify linguistic values with adjectives or adverbs. For example, we can use the hedges rather and somewhat to construct the additional values rather old or somewhat young.

Fuzzification operations can map mathematical input values into fuzzy membership functions. And the opposite de-fuzzifying operations can be used to map a fuzzy output membership function into a "crisp" output value that can be then used for decision or control purposes.

Process

- Fuzzify all input values into fuzzy membership functions.
- Execute all applicable rules in the rulebase to compute the fuzzy output functions.
- De-fuzzify the fuzzy output functions to get "crisp" output values.

Fuzzification

Fuzzification is the process of assigning the numerical input of a system to fuzzy sets with some degree of membership. This degree of membership may be anywhere within the interval [0,1]. If it is 0 then the value does not belong to the given fuzzy set, and if it is 1 then the value completely belongs within the fuzzy set. Any value between 0 and 1 represents the degree of uncertainty that the value belongs in the set. These fuzzy sets are typically described by words, and so by assigning the system input to fuzzy sets, we can reason with it in a linguistically natural manner.

For example, in the image below the meanings of the expressions cold, warm, and hot are represented by functions mapping a temperature scale. A point on that scale has three "truth values"—one for each of the three functions. The vertical line in the image represents a particular temperature that the three arrows (truth values) gauge. Since the red arrow points to zero, this temperature may be interpreted as "not hot"; i.e. this temperature has zero membership in the fuzzy set "hot". The orange arrow (pointing at 0.2) may describe it as "slightly warm" and the blue arrow (pointing at 0.8) "fairly cold". Therefore, this temperature has 0.2 memberships in the fuzzy set "warm" and 0.8 memberships in the fuzzy set "cold". The degree of membership assigned for each fuzzy set is the result of fuzzification.

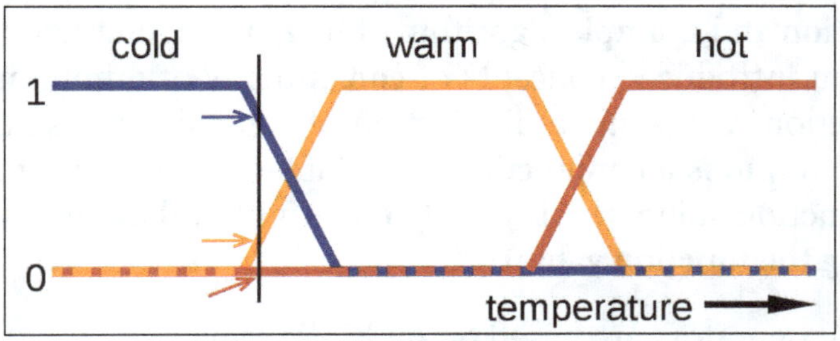

Fuzzy logic temperature.

Fuzzy sets are often defined as triangle or trapezoid-shaped curves, as each value will have a slope where the value is increasing, a peak where the value is equal to 1 (which can have a length of 0 or greater) and a slope where the value is decreasing. They can also be defined using a sigmoid function. One common case is the standard logistic function defined as:

$$S(x) = \frac{1}{1+e^{-x}}$$

which has the following symmetry property:

$$S(x) + S(-x) = 1$$

from this it follows that:

$$(S(x)+S(-x)) \cdot (S(y)+S(-y)) \cdot (S(z)+S(-z)) = 1$$

Fuzzy Logic Operators

Fuzzy logic works with membership values in a way that mimics Boolean logic. To this end, replacements for basic operators AND, OR, NOT must be available. There are several ways to this. A common replacement is called the Zadeh operators:

Boolean	Fuzzy
AND(x, y)	MIN(x, y)
OR(x, y)	MAX(x, y)
NOT(x)	1 − x

For TRUE/1 and FALSE/0, the fuzzy expressions produce the same result as the Boolean expressions.

There are also other operators, more linguistic in nature, called hedges that can be applied. These are generally adverbs such as very, or somewhat, which modify the meaning of a set using a mathematical formula.

However, an arbitrary choice table does not always define a fuzzy logic function. In the paper, a criterion has been formulated to recognize whether a given choice table defines a fuzzy logic function and a simple algorithm of fuzzy logic function synthesis has been proposed based on introduced concepts of constituents of minimum and maximum. A fuzzy logic function represents a disjunction of constituents of minimum, where a constituent of minimum is a conjunction of variables of the current area greater than or equal to the function value in this area (to the right of the function value in the inequality, including the function value).

Another set of And operators are based on multiplication:

```
x AND y = x*y

x OR y = 1-(1-x)*(1-y)  = x+y-x*y
```

1−(1−x)*(1−y) comes from this:

```
x OR y = NOT( AND( NOT(x), NOT(y) ) )

x OR y = NOT( AND(1-x, 1-y) )

x OR y = NOT( (1-x)*(1-y) )

x OR y = 1-(1-x)*(1-y)
```

IF-THEN Rules

IF-THEN rules map input or computed truth values to desired output truth values. Example:

```
IF temperature IS very cold THEN fan_speed is stopped

IF temperature IS cold THEN fan_speed is slow

IF temperature IS warm THEN fan_speed is moderate

IF temperature IS hot THEN fan_speed is high
```

Given a certain temperature, the fuzzy variable hot has a certain truth value, which is copied to the high variable.

Should an output variable occur in several THEN parts, then the values from the respective IF parts are combined using the OR operator.

Defuzzification

The goal is to get a continuous variable from fuzzy truth values.

This would be easy if the output truth values were exactly those obtained from fuzzification of a given number. Since, however, all output truth values are computed independently; in most cases they do not represent such a set of numbers. One has then to decide for a number that matches best the "intention" encoded in the truth value. For example, for several truth values of fan_speed, an actual speed must be found that best fits the computed truth values of the variables 'slow', 'moderate' and so on.

There is no single algorithm for this purpose.

A common algorithm is:

- For each truth value, cut the membership function at this value.
- Combine the resulting curves using the OR operator.
- Find the center-of-weight of the area under the curve.
- The x position of this center is then the final output.

Forming a Consensus of Inputs and Fuzzy Rules

Since the fuzzy system output is a consensus of all of the inputs and all of the rules, fuzzy logic systems can be well behaved when input values are not available or are not trustworthy. Weightings can be optionally added to each rule in the rulebase and weightings can be used to regulate the degree to which a rule affects the output values. These rule weightings can be based upon the priority, reliability or consistency of each rule. These rule weightings may be static or can be changed dynamically, even based upon the output from other rules.

Early Applications

Many of the early successful applications of fuzzy logic were implemented in Japan. The first notable application was on the subway train in Sendai, in which fuzzy logic was able to improve the economy, comfort, and precision of the ride. It has also been used in recognition of hand-written symbols in Sony pocket computers, flight aid for helicopters, controlling of subway systems in order to improve driving comfort, precision of halting, and power economy, improved fuel consumption for automobiles, single-button control for washing machines, automatic motor control for vacuum cleaners with recognition of surface condition and degree of soiling, and prediction systems for early recognition of earthquakes through the Institute of Seismology Bureau of Meteorology, Japan.

Current Applications

In Medical Decision Making

Fuzzy logic is an important concept when it comes to medical decision making. Since medical and healthcare data can be subjective or fuzzy, applications in this domain have a great potential to benefit a lot by using fuzzy logic based approaches. One of the common application areas that use fuzzy logic is computer-aided diagnosis (CAD) in medicine. CAD is a computerized set of inter-related tools which can be used to aid physicians in their diagnostic decision-making. For example, when a physician finds a lesion which is abnormal but still at a very early stage of development he/she may use a CAD approach to characterize the lesion and diagnose its nature. Fuzzy logic can be highly appropriate to describe key characteristics of this lesion. Fuzzy logic can be used in many different aspects within the CAD framework. Such aspects include in medical image analysis, biomedical signal analysis, segmentation of images or signals, and feature extraction/selection of images or signals as described.

The biggest question in this application area is how much useful information can be derived when using fuzzy logic. A major challenge is how to derive the required fuzzy data. This is even more challenging when one has to elicit such data from humans (usually, patients). As it said "The envelope of what can be achieved and what cannot be achieved in medical diagnosis, ironically, is itself a fuzzy one". How to elicit fuzzy data, and how to validate the accuracy of the data is still an ongoing effort strongly related to the application of fuzzy logic. The problem of assessing the quality of fuzzy data is a difficult one. This is why fuzzy logic is a highly promising possibility within the CAD application area but still requires more research to achieve its full potential. Although the concepts of using fuzzy logic in CAD is exciting, there are still several challenges that fuzzy approaches face within the CAD framework.

Logical Analysis

In mathematical logic, there are several formal systems of "fuzzy logic", most of which are in the family of t-norm fuzzy logics.

Propositional Fuzzy Logics

The most important propositional fuzzy logics are:

- Monoidal t-norm-based propositional fuzzy logic MTL is an axiomatization of logic where conjunction is defined by a left continuous t-norm and implication is defined as the residuum of the t-norm. Its models correspond to MTL-algebras that are pre-linear commutative bounded integral residuated lattices.
- Basic propositional fuzzy logic BL is an extension of MTL logic where conjunction

is defined by a continuous t-norm, and implication is also defined as the residuum of the t-norm. Its models correspond to BL-algebras.

- Łukasiewicz fuzzy logic is the extension of basic fuzzy logic BL where standard conjunction is the Łukasiewicz t-norm. It has the axioms of basic fuzzy logic plus an axiom of double negation, and its models correspond to MV-algebras.
- Gödel fuzzy logic is the extension of basic fuzzy logic BL where conjunction is Gödel t-norm. It has the axioms of BL plus an axiom of idempotence of conjunction, and its models are called G-algebras.
- Product fuzzy logic is the extension of basic fuzzy logic BL where conjunction is product t-norm. It has the axioms of BL plus another axiom for cancellativity of conjunction, and its models are called product algebras.
- Fuzzy logic with evaluated syntax (sometimes also called Pavelka's logic), denoted by EVŁ, is a further generalization of mathematical fuzzy logic. While the above kinds of fuzzy logic have traditional syntax and many-valued semantics, in EVŁ is evaluated also syntax. This means that each formula has an evaluation. Axiomatization of EVŁ stems from Łukasziewicz fuzzy logic. A generalization of classical Gödel completeness theorem is provable in EVŁ^.

Predicate Fuzzy Logics

These extend the above-mentioned fuzzy logics by adding universal and existential quantifiers in a manner similar to the way that predicate logic is created from propositional logic. The semantics of the universal (resp. existential) quantifier in t-norm fuzzy logics is the infimum (resp. supremum) of the truth degrees of the instances of the quantified subformula.

Decidability Issues for Fuzzy Logic

The notions of a "decidable subset" and "recursively enumerable subset" are basic ones for classical mathematics and classical logic. Thus the question of a suitable extension of them to fuzzy set theory is a crucial one. A first proposal in such a direction was made by E.S. Santos by the notions of fuzzy Turing machine; Markov normal fuzzy algorithm and fuzzy program. Successively, L. Biacino and G. Gerla argued that the proposed definitions are rather questionable. For example, one shows that the fuzzy Turing machines are not adequate for fuzzy language theory since there are natural fuzzy languages intuitively computable that cannot be recognized by a fuzzy Turing Machine. Then, they proposed the following definitions. Denote by Ü the set of rational numbers in [0,1]. Then a fuzzy subset s : S → [0,1] of a set S is recursively enumerable if a recursive map h : S×N → Ü exists such that, for every x in S, the function h(x, n) is increasing with respect to n and s(x) = lim h(x, n). We say that s is decidable if both s and its complement –s are recursively enumerable. An extension of such a theory to the general case of the L-subsets is possible. The proposed definitions are well related with fuzzy logic. Indeed, the following theorem holds true (provided that the deduction apparatus of the considered fuzzy logic satisfies some obvious effectiveness property).

Any "axiomatizable" fuzzy theory is recursively enumerable. In particular, the fuzzy set of logically true formulas is recursively enumerable in spite of the fact that the crisp set of valid formulas is not recursively enumerable, in general. Moreover, any axiomatizable and complete theory is decidable.

It is an open question to give supports for a "Church thesis" for fuzzy mathematics, the proposed notion of recursive enumerability for fuzzy subsets is the adequate one. In order to solve this, an extension of the notions of fuzzy grammar and fuzzy Turing machine are necessary. Another open question is to start from this notion to find an extension of Gödel's theorems to fuzzy logic.

Fuzzy Databases

Once fuzzy relations are defined, it is possible to develop fuzzy relational databases. The first fuzzy relational database, FRDB, appeared in Maria Zemankova's dissertation. Later, some other models arose like the Buckles-Petry model, the Prade-Testemale Model, the Umano-Fukami model or the GEFRED model by J.M. Medina, M.A. Vila et al.

Fuzzy querying languages have been defined, such as the SQLf by P. Bosc et al. and the FSQL by J. Galindo et al. These languages define some structures in order to include fuzzy aspects in the SQL statements, like fuzzy conditions, fuzzy comparators, fuzzy constants, fuzzy constraints, fuzzy thresholds, linguistic labels etc.

The knowledge graph Weaviate uses fuzzy logic to index data based on a machine learning model called the Contextionary.

Comparison to Probability

Fuzzy logic and probability address different forms of uncertainty. While both fuzzy logic and probability theory can represent degrees of certain kinds of subjective belief, fuzzy set theory uses the concept of fuzzy set membership, i.e., how much an observation is within a vaguely defined set, and probability theory uses the concept of subjective probability, i.e., frequency of occurrence or likelihood of some event or condition. The concept of fuzzy sets was developed in the mid-twentieth century at Berkeley as a response to the lacking of probability theory for jointly modelling uncertainty and vagueness.

Bart Kosko claims in Fuzziness vs. Probability that probability theory is a subtheory of fuzzy logic, as questions of degrees of belief in mutually-exclusive set membership in probability theory can be represented as certain cases of non-mutually-exclusive graded membership in fuzzy theory. In that context, he also derives Bayes' theorem from the concept of fuzzy subsethood. Lotfi A. Zadeh argues that fuzzy logic is different in character from probability, and is not a replacement for it. He fuzzified probability to fuzzy probability and also generalized it to possibility theory.

More generally, fuzzy logic is one of many different extensions to classical logic intended to deal with issues of uncertainty outside of the scope of classical logic, the inapplicability of probability theory in many domains, and the paradoxes of Dempster-Shafer theory.

Relation to Ecorithms

Computational theorist Leslie Valiant uses the term ecorithms to describe how many less exact systems and techniques like fuzzy logic (and "less robust" logic) can be applied to learning algorithms. Valiant essentially redefines machine learning as evolutionary. In general use, ecorithms are algorithms that learn from their more complex environments (hence eco-) to generalize, approximate and simplify solution logic. Like fuzzy logic, they are methods used to overcome continuous variables or systems too complex to completely enumerate or understand discretely or exactly. Ecorithms and fuzzy logic also have the common property of dealing with possibilities more than probabilities, although feedback and feed forward, basically stochastic weights, are a feature of both when dealing with, for example, dynamical systems.

Compensatory Fuzzy Logic

Compensatory fuzzy logic (CFL) is a branch of fuzzy logic with modified rules for conjunction and disjunction. When the truth value of one component of a conjunction or disjunction is increased or decreased, the other component is decreased or increased to compensate. This increase or decrease in truth value may be offset by the increase or decrease in another component. An offset may be blocked when certain thresholds are met. Proponents claim that CFL allows for better computational semantic behaviors and mimic natural language.

Compensatory Fuzzy Logic consists of four continuous operators: conjunction (c); disjunction (d); fuzzy strict order (or); and negation (n). The conjunction is the geometric mean and its dual as conjunctive and disjunctive operators.

Inner Model Theory

In set theory, inner model theory is the study of certain models of ZFC or some fragment or strengthening thereof. Ordinarily these models are transitive subsets or subclasses of the von Neumann universe V, or sometimes of a generic extension of V. Inner model theory studies the relationships of these models to determinacy, large cardinals, and descriptive set theory. Despite the name, it is considered more a branch of set theory than of model theory.

Examples:

- The class of all sets is an inner model containing all other inner models.
- The first non-trivial example of an inner model was the constructible universe L developed by Kurt Gödel. Every model M of ZF has an inner model L^M satisfying the axiom of constructibility, and this will be the smallest inner model of M containing all the ordinals of M. Regardless of the properties of the original model, L^M will satisfy the generalized continuum hypothesis and combinatorial axioms such as the diamond principle $\diamond$.
- The sets that are hereditarily ordinal definable form an inner model.
- The sets that are hereditarily definable over a countable sequence of ordinals form an inner model, used in Solovay's theorem.
- L(R).
- L[U].

Consistency Results

One important use of inner models is the proof of consistency results. If it can be shown that every model of an axiom A has an inner model satisfying axiom B, then if A is consistent, B must also be consistent. This analysis is most useful when A is an axiom independent of ZFC, for example a large cardinal axiom; it is one of the tools used to rank axioms by consistency strength.

Large Cardinal

In the mathematical field of set theory, a large cardinal property is a certain kind of property of transfinite cardinal numbers. Cardinals with such properties are, as the name suggests, generally very "large" (for example, bigger than the least α such that $\alpha=\omega_\alpha$). The proposition that such cardinals exist cannot be proved in the most common axiomatization of set theory, namely ZFC, and such propositions can be viewed as ways of measuring how "much", beyond ZFC, one needs to assume to be able to prove certain desired results. In other words, they can be seen, in Dana Scott's phrase, as quantifying the fact "that if you want more you have to assume more".

There is a rough convention that results provable from ZFC alone may be stated without hypotheses, but that if the proof requires other assumptions (such as the existence of large cardinals), these should be stated. Whether this is simply a linguistic convention, or something more, is a controversial point among distinct philosophical schools.

A large cardinal axiom is an axiom stating that there exists a cardinal (or perhaps many of them) with some specified large cardinal property.

Most working set theorists believe that the large cardinal axioms that are currently being considered are consistent with ZFC. These axioms are strong enough to imply the consistency of ZFC. This has the consequence (via Gödel's second incompleteness theorem) that their consistency with ZFC cannot be proven in ZFC (assuming ZFC is consistent).

There is no generally agreed precise definition of what a large cardinal property is, though essentially everyone agrees that those in the list of large cardinal properties are large cardinal properties.

A necessary condition for a property of cardinal numbers to be a large cardinal property is that the existence of such a cardinal is not known to be inconsistent with ZFC and it has been proven that if ZFC is consistent, then ZFC + "no such cardinal exists" is consistent.

Hierarchy of Consistency Strength

A remarkable observation about large cardinal axioms is that they appear to occur in strict linear order by consistency strength. That is, no exception is known to the following: Given two large cardinal axioms A1 and A2, exactly one of three things happens:

ZFC proves "ZFC+A1 is consistent if and only if ZFC+A2 is consistent":

- ZFC+A1 proves that ZFC+A2 is consistent.
- ZFC+A2 proves that ZFC+A1 is consistent.

These are mutually exclusive, unless one of the theories in question is actually inconsistent.

In case 1 we say that A1 and A2 are equiconsistent. In case 2, we say that A1 is consistency-wise stronger than A2. If A2 is stronger than A1, then ZFC+A1 cannot prove ZFC+A2 is consistent, even with the additional hypothesis that ZFC+A1 is itself consistent (provided of course that it really is). This follows from Gödel's second incompleteness theorem.

The observation that large cardinal axioms are linearly ordered by consistency strength is just that, an observation, not a theorem. (Without an accepted definition of large cardinal property, it is not subject to proof in the ordinary sense). Also, it is not known in every case which of the three cases holds. Saharon Shelah has asked, "is there some theorem explaining this, or is our vision just more uniform than we realize?" Woodin, however, deduces this from the Ω-conjecture, the main unsolved problem of his Ω-logic. It is also noteworthy that many combinatorial statements are exactly equiconsistent with some large cardinal rather than, say, being intermediate between them.

The order of consistency strength is not necessarily the same as the order of the size of the smallest witness to a large cardinal axiom. For example, the existence of a huge cardinal is much stronger, in terms of consistency strength, than the existence of a supercompact cardinal, but assuming both exist, the first huge is smaller than the first supercompact.

Motivations and Epistemic Status

Large cardinals are understood in the context of the von Neumann universe V, which is built up by transfinitely iterating the powerset operation, which collects together all subsets of a given set. Typically, models in which large cardinal axioms fail can be seen in some natural way as submodels of those in which the axioms hold. For example, if there is an inaccessible cardinal, then "cutting the universe off" at the height of the first such cardinal yields a universe in which there is no inaccessible cardinal. Or if there is a measurable cardinal, then iterating the definable powerset operation rather than the full one yields Gödel's constructible universe, L, which does not satisfy the statement "there is a measurable cardinal" (even though it contains the measurable cardinal as an ordinal).

Thus, from a certain point of view held by many set theorists (especially those inspired by the tradition of the Cabal), large cardinal axioms "say" that we are considering all the sets we're "supposed" to be considering, whereas their negations are "restrictive" and say that we're considering only some of those sets. Moreover the consequences of large cardinal axioms seem to fall into natural patterns (see Maddy, "Believing the Axioms, II"). For these reasons, such set theorists tend to consider large cardinal axioms to have a preferred status among extensions of ZFC, one not shared by axioms of less clear motivation (such as Martin's axiom) or others that they consider intuitively unlikely (such as V = L). The hardcore realists in this group would state, more simply, that large cardinal axioms are true.

This point of view is by no means universal among set theorists. Some formalists would assert that standard set theory is by definition the study of the consequences of ZFC, and while they might not be opposed in principle to studying the consequences of other systems, they see no reason to single out large cardinals as preferred. There are also realists who deny that ontological maximalism is a proper motivation, and even believe that large cardinal axioms are false. And finally, there are some who deny that the negations of large cardinal axioms are restrictive, pointing out that (for example) there can be a transitive set model in L that believes there exists a measurable cardinal, even though L itself does not satisfy that proposition.

Measurable Cardinal

In mathematics, a measurable cardinal is a certain kind of large cardinal number. In order to define the concept, one introduces a two-valued measure on a cardinal κ, or more generally on any set. For a cardinal κ, it can be described as a subdivision of all of its subsets into large and small sets such that κ itself is large, $\emptyset$ and all singletons $\{\alpha\}$, $\alpha \in \kappa$ are small, complements of small sets are large and vice versa. The intersection of fewer than κ large sets is again large.

It turns out that uncountable cardinals endowed with a two-valued measure are large cardinals whose existence cannot be proved from ZFC.

The concept of a measurable cardinal was introduced by Stanislaw Ulam in 1930.

Formally, a measurable cardinal is an uncountable cardinal number κ such that there exists a κ-additive, non-trivial, 0-1-valued measure on the power set of κ. (Here the term κ-additive means that, for any sequence A_α, $\alpha < \lambda$ of cardinality $\lambda < \kappa$, A_α being pairwise disjoint sets of ordinals less than κ, the measure of the union of the A_α equals the sum of the measures of the individual A_α.)

Equivalently, κ is measurable means that it is the critical point of a non-trivial elementary embedding of the universe V into a transitive class M. This equivalence is due to Jerome Keisler and Dana Scott, and uses the ultrapower construction from model theory. Since V is a proper class, a technical problem that is not usually present when considering ultrapowers needs to be addressed, by what is now called Scott's trick.

Equivalently, κ is a measurable cardinal if and only if it is an uncountable cardinal with a κ-complete, non-principal ultrafilter. Again, this means that the intersection of any strictly less than κ-many sets in the ultrafilter, is also in the ultrafilter.

Properties

Although it follows from ZFC that every measurable cardinal is inaccessible (and is ineffable, Ramsey, etc.), it is consistent with ZF that a measurable cardinal can be a successor cardinal. It follows from ZF + axiom of determinacy that ω_1 is measurable, and that every subset of ω_1 contains or is disjoint from a closed and unbounded subset.

Ulam showed that the smallest cardinal κ that admits a non-trivial countably-additive two-valued measure must in fact admit a κ-additive measure. (If there were some collection of fewer than κ measure-0 subsets whose union was κ, then the induced measure on this collection would be a counterexample to the minimality of κ.) From there, one can prove (with the Axiom of Choice) that the least such cardinal must be inaccessible.

It is trivial to note that if κ admits a non-trivial κ-additive measure, then κ must be regular. (By non-triviality and κ-additivity, any subset of cardinality less than κ must have measure 0, and then by κ-additivity again, this means that the entire set must not be a union of fewer than κ sets of cardinality less than κ.) Finally, if $\lambda < \kappa$, then it can't be the case that $\kappa \leq 2^{\lambda}$. If this were the case, then we could identify κ with some collection of 0-1 sequences of length λ. For each position in the sequence, either the subset of sequences with 1 in that position or the subset with 0 in that position would have to have measure 1. The intersection of these λ-many measure 1 subsets would thus also have to have measure 1, but it would contain exactly one sequence, which would contradict the non-triviality of the measure. Thus, assuming the Axiom of Choice, we can infer that κ is a strong limit cardinal, which completes the proof of its inaccessibility.

If κ is measurable and $p \in V_\kappa$ and M (the ultrapower of V) satisfies $\psi(\kappa, p)$, then the set of $\alpha < \kappa$ such that V satisfies $\psi(\alpha, p)$ is stationary in κ (actually a set of measure 1). In particular if ψ is a Π_1 formula and V satisfies $\psi(\kappa, p)$, then M satisfies it and thus V satisfies $\psi(\alpha,p)$ for a stationary set of $\alpha < \kappa$. This property can be used to show that κ is a limit of most types of large cardinals that are weaker than measurable. Notice that the ultrafilter or measure witnessing that κ is measurable cannot be in M since the smallest such measurable cardinal would have to have another such below it, which is impossible.

If one starts with an elementary embedding j_1 of V into M_1 with critical point κ, then one can define an ultrafilter U on κ as $\{ S \subseteq \kappa : \kappa \in j_1(S) \}$. Then taking an ultrapower of V over U we can get another elementary embedding j_2 of V into M_2. However, it is important to remember that $j_2 \neq j_1$. Thus other types of large cardinals such as strong cardinals may also be measurable, but not using the same embedding. It can be shown that a strong cardinal κ is measurable and also has κ-many measurable cardinals below it.

Every measurable cardinal κ is a 0-huge cardinal because ${}^{\kappa}M \subseteq M$, that is, every function from κ to M is in M. Consequently, $V_{\kappa+1} \subseteq M$.

Real-valued Measurable

A cardinal κ is called real-valued measurable if there is a κ-additive probability measure on the power set of κ that vanishes on singletons. Real-valued measurable cardinals were introduced by Stefan Banach. Banach and Kuratowski showed that the continuum hypothesis implies that $\mathfrak{c}$ is not real-valued measurable. Stanislaw Ulam showed that real valued measurable cardinals are weakly inaccessible (they are in fact weakly Mahlo). All measurable cardinals are real-valued measurable, and a real-valued measurable cardinal κ is measurable if and only if κ is greater than $\mathfrak{c}$. Thus a cardinal is measurable if and only if it is real-valued measurable and strongly inaccessible. A real valued measurable cardinal less than or equal to $\mathfrak{c}$ exists if and only if there is a countably additive extension of the Lebesgue measure to all sets of real numbers if and only if there is an atomless probability measure on the power set of some non-empty set.

Solovay showed that existence of measurable cardinals in ZFC, real valued measurable cardinals in ZFC, and measurable cardinals in ZF, are equiconsistent.

Weak Inaccessibility of Real-valued Measurable Cardinals

Say that a cardinal number α is an Ulam number if, whenever:

- μ is an outer measure on a set X.
- $\mu(X) < \infty$.
- $\mu(\{x\}) = 0,\ x \in X$.
- All $A \subset X$ are μ-measurable.

Then,

$$\operatorname{card} X \leq \alpha \Rightarrow \mu(X) = 0$$

Equivalently, a cardinal number α is an Ulam number if, whenever:

- ν is an outer measure on a set Y, and F a disjoint family of subsets of Y.
- $\nu\left(\bigcup F\right) < \infty$.
- $\nu(A) = 0$ for $A \in F$.
- $\bigcup G$ is ν-measurable for every $G \subset F$.

Then,

$$\operatorname{card} F \leq \alpha \Rightarrow \nu\left(\bigcup F\right) = 0$$

The smallest infinite cardinal $\aleph_0$ is an Ulam number. The class of Ulam numbers is closed under the cardinal successor operation. If an infinite cardinal β has an immediate predecessor α that is an Ulam number, assume μ satisfies properties (1)–(4) with X = β. In the von Neumann model of ordinals and cardinals, choose injective functions:

$$f_x : x \to \alpha, \quad \forall x \in \beta$$

And define the sets:

$$U(b,a) = \{x \in \beta : f_x(b) = a\}, \quad a \in \alpha, b \in \beta$$

Since the f_x are one-to-one, the sets:

$$\{U(b,a), b \in \beta\}\text{(a fixed)}$$
$$\{U(b,a), a \in \alpha\}\text{(b fixed)}$$

are disjoint. By property ($\nu\left(\bigcup F\right)<\infty$) of μ, the set:

$$\{b \in \beta : \mu(U(b,a)) > o\}$$

is countable, and hence:

$$\text{card}\{(b,a) \in \beta \times \alpha \mid \mu(U(b,a)) > o\} \leq \aleph_o \cdot \alpha = \alpha$$

Thus there is a b_o such that:

$$\mu(U(b_o,a)) = o \quad \forall a \in \alpha$$

Implying, since α is an Ulam number and using the second definition v = μ and conditions (previous four conditions) fulfilled),

$$\mu\left(\bigcup_{a\in\alpha} U(b_o,a)\right) = o$$

If $b_o < x < \beta$, then $f_x(b_o) = a_x \Rightarrow x \in U(b_o,a_x)$. Thus,

$$\beta = b_o \cup \{b_o\} \cup \bigcup_{a\in\alpha} U(b_o,a)$$

By property ($\nu\left(\bigcup F\right)<\infty$), $\mu\{b_o\} = o$, and since $\text{card}\, b_o \leq \alpha$, by ($\bigcup G$), ($\nu\left(\bigcup F\right)<\infty$) and ($\nu(A) = o$), $\mu(b_o) = o$. It follows that $\mu(\beta) = o$. The conclusion is that β is an Ulam number. There is a similar proof that the supremum of a set S of Ulam numbers with card S an Ulam number is again a Ulam number. Together with the previous result, this imples that a cardinal that is not an Ulam number is weakly inaccessible.

Woodin Cardinal

In set theory, a Woodin cardinal is a cardinal number λ such that for all functions:

$$f : \lambda \to \lambda$$

There exists a cardinal κ < λ with:

$$\{f(\beta) \mid \beta < \kappa\} \subseteq \kappa$$

And an elementary embedding:

$$j : V \to M$$

From the Von Neumann universe V into a transitive inner model M with critical point κ and,

$$V_{j(f)(\kappa)} \subseteq M$$

An equivalent definition is this: λ is Woodin if and only if λ is strongly inaccessible and for all $A \subseteq V_\lambda$ there exists a $\lambda_A < \lambda$ which is < λ-A-strong.

λ_A being < λ-A-strong means that for all ordinals α < λ, there exist a $j: V \to M$ which is an elementary embedding with critical point λ_A, $j(\lambda_A) > \alpha$, $V_\alpha \subseteq M$ and $j(A) \cap V_\alpha = A \cap V_\alpha$.

A Woodin cardinal is preceded by a stationary set of measurable cardinals, and thus it is a Mahlo cardinal. However, the first Woodin cardinal is not even weakly compact.

Consequences

Woodin cardinals are important in descriptive set theory. By a result of Martin and Steel, existence of infinitely many Woodin cardinals implies projective determinacy, which in turn implies that every projective set is measurable, has the Baire property (differs from an open set by a meager set, that is, a set which is a countable union of nowhere dense sets), and the perfect set property (is either countable or contains a perfect subset).

The consistency of the existence of Woodin cardinals can be proved using determinacy hypotheses. Working in ZF+AD+DC one can prove that Θ_0 is Woodin in the class of hereditarily ordinal-definable sets. Θ_0 is the first ordinal onto which the continuum cannot be mapped by an ordinal-definable surjection.

Shelah proved that if the existence of a Woodin cardinal is consistent then it is consistent that the nonstationary ideal on ω_1 is $\aleph_2$-saturated. Woodin also proved the equiconsistency of the existence of infinitely many Woodin cardinals and the existence of an $\aleph_1$-dense ideal over $\aleph_1$.

Hyper-Woodin Cardinals

A cardinal κ is called hyper-Woodin if there exists a normal measure U on κ such that for every set S, the set:

$$\{\lambda < \kappa \mid \lambda \text{ is} < \kappa - S - \text{strong}\}$$

is in U.

λ is < κ-S-strong if and only if for each δ < κ there is a transitive class N and an elementary embedding:

$$j: V \to N$$

with,

$$\lambda = \text{crit}(j)$$

$$j(\lambda) \geq \delta$$

and,

$$j(S) \cap H_\delta = S \cap H_\delta$$

The name alludes to the classical result that a cardinal is Woodin if and only if for every set S, the set:

$$\{\lambda < \kappa \mid \lambda \text{ is } < \kappa - S - \text{strong}\}$$

is a stationary set.

The measure U will contain the set of all Shelah cardinals below κ.

Weakly Hyper-woodin Cardinals

A cardinal κ is called weakly hyper-Woodin if for every set S there exists a normal measure U on κ such that the set {λ < κ | λ is <κ-S-strong} is in U. λ is <κ-S-strong if and only if for each δ < κ there is a transitive class N and an elementary embedding j : V → N with λ = crit(j), j(λ) >= δ, and $j(S) \cap H_\delta = S \cap H_\delta$.

The name alludes to the classic result that a cardinal is Woodin if for every set S, the set {λ < κ | λ is <κ-S-strong} is stationary.

The difference between hyper-Woodin cardinals and weakly hyper-Woodin cardinals is that the choice of U does not depend on the choice of the set S for hyper-Woodin cardinals.

Cardinal Function

In mathematics, a cardinal function (or cardinal invariant) is a function that returns cardinal numbers.

Cardinal Functions in Set Theory

- The most frequently used cardinal function is a function which assigns to a set "A" its cardinality, denoted by |A|.
- Aleph numbers and Beth numbers can both be seen as cardinal functions defined on ordinal numbers.

- Cardinal arithmetic operations are examples of functions from cardinal numbers (or pairs of them) to cardinal numbers.
- Cardinal characteristics of a (proper) ideal I of subsets of X are:

$$\mathrm{add}(I) = \min\{|\mathcal{A}| : \mathcal{A} \subseteq I \wedge \bigcup \mathcal{A} \notin I\}$$

The "additivity" of I is the smallest number of sets from I whose union is not in I any more. As any ideal is closed under finite unions, this number is always at least $\aleph_0$; if I is a σ-ideal, then $\mathrm{add}(I) \geq \aleph_1$.

$$\mathrm{cov}(I) = \min\{|\mathcal{A}| : \mathcal{A} \subseteq I \wedge \bigcup \mathcal{A} = X\}$$

The "covering number" of I is the smallest number of sets from I whose union is all of X. As X itself is not in I, we must have $\mathrm{add}(I) \leq \mathrm{cov}(I)$.

$$\mathrm{non}(I) = \min\{|A| : A \subseteq X \wedge A \notin I\}$$

The "uniformity number" of I (sometimes also written unif(I) is the size of the smallest set not in I. Assuming I contains all singletons, $\mathrm{add}(I) \leq \mathrm{non}(I)$.

$$\mathrm{cof}(I) = \min\{|\mathcal{B}| : \mathcal{B} \subseteq I \wedge (\forall A \in I)(\exists B \in \mathcal{B})(A \subseteq B)\}$$

The "cofinality" of I is the cofinality of the partial order $(I, \subseteq)$. It is easy to see that we must have $\mathrm{non}(I) \leq \mathrm{cof}(I)$ and $\mathrm{cov}(I) \leq \mathrm{cof}(I)$.

In the case that I is an ideal closely related to the structure of the reals, such as the ideal of Lebesgue null sets or the ideal of meagre sets, these cardinal invariants are referred to as cardinal characteristics of the continuum.

- For a preordered set $(\mathbb{P}, \sqsubseteq)$ the bounding number $\mathfrak{b}(\mathbb{P})$ and dominating number $\mathfrak{d}(\mathbb{P})$ is defined as:

$$\mathfrak{b}(\mathbb{P}) = \min\{|Y| : Y \subseteq \mathbb{P} \wedge (\forall x \in \mathbb{P})(\exists y \in Y)(y \not\sqsubseteq x)\}$$
$$\mathfrak{d}(\mathbb{P}) = \min\{|Y| : Y \subseteq \mathbb{P} \wedge (\forall x \in \mathbb{P})(\exists y \in Y)(x \sqsubseteq y)\}$$

- In PCF theory the cardinal function $pp_\kappa(\lambda)$ is used.

Cardinal Functions in Topology

Cardinal functions are widely used in topology as a tool for describing various topological properties. Below are some examples. (Note: some authors, arguing that "there are no finite cardinal numbers in general topology", prefer to define the cardinal functions listed below so that they never taken on finite cardinal numbers as values; this requires modifying some of the definitions given below, e.g. by adding "+ℵ" to the right-hand side of the definitions, etc.)

- Perhaps the simplest cardinal invariants of a topological space X are its cardinality and the cardinality of its topology, denoted respectively by |X| and o(X).
- The weight w(X) of a topological space X is the cardinality of the smallest base for X. When $w(X) = \aleph_0$ the space X is said to be second countable:
    - The π-weight of a space X is the cardinality of the smallest π-base for X.
    - The network weight of X is the smallest cardinality of a network for X. A network is a family $\mathcal{N}$ of sets, for which, for all points x and open neighbourhoods U containing x, there exists B in $\mathcal{N}$ for which $x \in B \subseteq U$.
- The character of a topological space X at a point x is the cardinality of the smallest local base for x. The character of space X is:

$$\chi(X) = \sup\{\chi(x, X) : x \in X\}.$$

When $\chi(X) = \aleph_0$ the space X is said to be first countable.

- The density d(X) of a space X is the cardinality of the smallest dense subset of X. When $d(X)=\aleph_0$ the space X is said to be separable.
- The Lindelöf number L(X) of a space X is the smallest infinite cardinality such that every open cover has a subcover of cardinality no more than L(X). When $L(X)=\aleph_0$ the space X is said to be a Lindelöf space.
- The cellularity or Suslin number of a space X is:

$$c(X) = \sup\{|\mathcal{U}| : \mathcal{U}$$

is a family of mutually disjoint non-empty open subsets of X}.

- The hereditary cellularity (sometimes spread) is the least upper bound of cellularities of its subsets:

$$s(X) = hc(X) = \sup\{c(Y) : Y \subseteq X\}$$

or,

$$s(X) = \sup\{|Y| : Y \subseteq X \text{ with the subspace topology is discrete}\}.$$

- The tightness t(x, X) of a topological space X at a point $x \in X$ is the smallest cardinal number α such that, whenever $x \in cl_X(Y)$ for some subset Y of X, there exists a subset Z of Y, with |Z| ≤ α, such that $x \in cl_X(Z)$. Symbolically,

$$t(x, X) = \sup\{\min\{|Z| : Z \subseteq Y \wedge x \in cl_X(Z)\} : Y \subseteq X \wedge x \in cl_X(Y)\}$$

The tightness of a space X is $t(X)=\sup\{t(x,X):x\in X\}$. When $t(X)=\aleph_0$ the space X is said to be countably generated or countably tight.

- The augmented tightness of a space X, $t^+(X)$ is the smallest regular cardinal α such that for any $Y\subseteq X$, $x\in cl_X(Y)$ there is a subset Z of Y with cardinality less than α, such that $x\in cl_X(Z)$.

Basic Inequalities

$$c(X)\leq d(X)\ \leq w(X)\ \leq o(X)\ \leq\ 2^{|X|}$$
$$\chi(X)\leq w(X)$$
$$nw(X)\leq w(X) \text{ and } o(X)\ \leq\ 2^{nw(X)}$$

Cardinal Functions in Boolean Algebras

Cardinal functions are often used in the study of Boolean algebras. We can mention, for example, the following functions:

- Cellularity $c(\mathbb{B})$ of a Boolean algebra $\mathbb{B}$ is the supremum of the cardinalities of antichains in $\mathbb{B}$.
- Length length( ) of a Boolean algebra $\mathbb{B}$ is:
  $\text{length}(\mathbb{B})=\sup\{|A|:A\subseteq\mathbb{B}\text{ is a chain}\}$.
- Depth $\text{depth}(\mathbb{B})$ of a Boolean algebra $\mathbb{B}$ is:
  $\text{depth}(\mathbb{B})=\sup\{|A|:A\subseteq\mathbb{B}\text{ is a well-ordered subset}\}$.
- Incomparability $\text{Inc}(\mathbb{B})$ of a Boolean algebra $\mathbb{B}$ is:
  $\text{Inc}(\mathbb{B})=\sup\{|A|:A\subseteq\mathbb{B}\text{ such that } (\forall a,b\in A)(a\neq b\Rightarrow\neg(a\leq b\vee b\leq a))\}$
- Pseudo-weight $\pi(\mathbb{B})$ of a Boolean algebra $\mathbb{B}$ is:
  $\pi(\mathbb{B})=\min\{|A|:A\subseteq\mathbb{B}\setminus\{0\}\text{ such that } (\forall b\in B\setminus\{0\})(\exists a\in A)(a\leq b)\}$.

Cardinal Functions in Algebra

Examples of cardinal functions in algebra are:

- Index of a subgroup H of G is the number of cosets.
- Dimension of a vector space V over a field K is the cardinality of any Hamel basis of V.

- More generally, for a free module M over a ring R we define rank rank(M) as the cardinality of any basis of this module.
- For a linear subspace W of a vector space V we define codimension of W (with respect to V).
- For any algebraic structure it is possible to consider the minimal cardinality of generators of the structure.
- For algebraic extensions algebraic degree and separable degree are often employed (note that the algebraic degree equals the dimension of the extension as a vector space over the smaller field).
- For non-algebraic field extensions transcendence degree is likewise used.

Determinacy

Determinacy is a subfield of set theory, a branch of mathematics, which examines the conditions under which one or the other player of a game has a winning strategy, and the consequences of the existence of such strategies. Alternatively and similarly, "determinacy" is the property of a game whereby such a strategy exists.

The games studied in set theory are usually Gale–Stewart games—two-player games of perfect information in which the players make an infinite sequence of moves and there are no draws. The field of game theory studies more general kinds of games, including games with draws such as tic-tac-toe, chess, or infinite chess, or games with imperfect information such as poker.

Basic Notions

Games

The first sort of game we shall consider is the two-player game of perfect information of length ω, in which the players play natural numbers. These games are often called Gale–Stewart games.

In this sort of game there are two players, often named I and II, who take turns playing natural numbers, with I going first. They play "forever"; that is, their plays are indexed by the natural numbers. When they're finished, a predetermined condition decides which player won. This condition need not be specified by any definable rule; it may simply be an arbitrary (infinitely long) lookup table saying who has won given a particular sequence of plays.

More formally, consider a subset A of Baire space; recall that the latter consists of all ω-sequences of natural numbers. Then in the game GA, I play a natural number a_0, then II plays a_1, then I plays a_2, and so on. Then I win the game if and only if:

$$\langle a_0, a_1, a_2, \ldots \rangle \in A$$

And otherwise II wins. A is then called the payoff set of GA.

It is assumed that each player can see all moves preceding each of his moves, and also knows the winning condition.

Strategies

Informally, a strategy for a player is a way of playing in which his plays are entirely determined by the foregoing plays. Again, such a "way" does not have to be capable of being captured by any explicable "rule", but may simply be a lookup table.

More formally, a strategy for player I (for a game in the sense of the preceding subsection) is a function that accepts as an argument any finite sequence of natural numbers, of even length, and returns a natural number. If σ is such a strategy and $<a_0,\ldots,a_{2n-1}>$ is a sequence of plays, then $\sigma(<a_0,\ldots,a_{2n-1}>)$ is the next play I will make, if I is following the strategy σ. Strategies for II are just the same, substituting "odd" for "even".

Note that we have said nothing, as yet, about whether a strategy is in any way good. A strategy might direct a player to make aggressively bad moves, and it would still be a strategy. In fact it is not necessary even to know the winning condition for a game, to know what strategies exist for the game.

Winning Strategies

A strategy is winning if the player following it must necessarily win, no matter what his opponent plays. For example, if σ is a strategy for I, then σ is a winning strategy for I in the game GA if, for any sequence of natural numbers to be played by II, say $<a_1,a_3,a_5,\ldots,$ etc.>, the sequence of plays produced by σ when II plays thus, namely:

$$\langle \sigma(\langle\rangle), a_1, \sigma(\langle \sigma(\langle\rangle), a_1 \rangle), a_3, \ldots \rangle$$

is an element of A.

Determined Games

A (class of) game(s) is determined if for all instances of the game there is a winning strategy for one of the players (not necessarily the same player for each instance). Note that there cannot be a winning strategy for both players for the same game, for if there were, the two strategies could be played against each other. The resulting outcome would then, by hypothesis, be a win for both players, which is impossible.

Determinacy from Elementary Considerations

All finite games of perfect information in which draws do not occur are determined.

Real-world games of perfect information, such as tic-tac-toe, chess, or infinite chess, are always finished in a finite number of moves (in chess-games this assumes the 50-move rule is applied). If such a game is modified so that a particular player wins under any condition where the game would have been called a draw, then it is always determined. The condition that the game is always over (i.e. all possible extensions of the finite position result in a win for the same player) in a finite number of moves corresponds to the topological condition that the set A giving the winning condition for G_A is clopen in the topology of Baire space.

For example, modifying the rules of chess to make drawn games a win for Black makes chess a determined game. As it happens, chess has a finite number of positions and a draw-by-repetition rules, so with these modified rules, if play continues long enough without White having won, then Black can eventually force a win (due to the modification of draw = win for black).

The proof that such games are determined is rather simple: Player I simply plays not to lose; that is, player I plays to make sure that player II does not have a winning strategy after I's move. If player I cannot do this, then it means player II had a winning strategy from the beginning. On the other hand, if player I can play in this way, then I must win, because the game will be over after some finite number of moves, and player I can't have lost at that point.

This proof does not actually require that the game always be over in a finite number of moves, only that it be over in a finite number of moves whenever II wins. That condition, topologically, is that the set A is closed. This fact—that all closed games are determined—is called the Gale–Stewart theorem. Note that by symmetry, all open games are determined as well. (A game is open if I can win only by winning in a finite number of moves.)

Determinacy from ZFC

David Gale and F. M. Stewart proved the open and closed games are determined. Determinacy for second level of the Borel hierarchy games was shown by Wolfe in 1955. Over the following 20 years, additional research using ever-more-complicated arguments established that third and fourth levels of the Borel hierarchy are determined.

In 1975, Donald A. Martin proved that all Borel games are determined; that is, if A is a Borel subset of Baire space, then GA is determined. This result, known as Borel determinacy, is the best possible determinacy result provable in ZFC, in the sense that the determinacy of the next higher Wadge class is not provable in ZFC.

In 1971, before Martin obtained his proof, Harvey Friedman showed that any proof of Borel determinacy must use the axiom of replacement in an essential way, in order to iterate the powerset axiom transfinitely often. Friedman's work gives a level-by-level result detailing how many iterations of the powerset axiom are necessary to guarantee determinacy at each level of the Borel hierarchy.

For every integer n, ZFC\P proves determinacy in the nth level of the difference hierarchy of Π^0_3 sets, but ZFC\P does not prove that for every integer n nth level of the difference hierarchy of Π^0_3 sets is determined.

Measurable Cardinals

It follows from the existence of a measurable cardinal that every analytic game (also called a Σ^1_1 game) is determined, or equivalently that every coanalytic (or Π^1_1) game is determined.

Actually a measurable cardinal is more than enough. A weaker principle — the existence of $0^\#$ is sufficient to prove coanalytic determinacy, and a little bit more: The precise result is that the existence of $0\#$ is equivalent to the determinacy of all levels of the difference hierarchy below the ω^2 level, i.e. $\omega{\cdot}n$-Π^1_1 determinacy for every n.

From a measurable cardinal we can improve this very slightly to ω^2-Π^1_1 determinacy. From the existence of more measurable cardinals, one can prove the determinacy of more levels of the difference hierarchy over Π^1_1.

Proof of Determinacy from Sharps

For every real number r, Σ^1_1 determinacy is equivalent to existence of $r^\#$. To illustrate how large cardinals lead to determinacy, here is a proof of $\Sigma^1_1(r)$ determinacy given existence of $r^\#$.

Let A be a $\Sigma^1_1(r)$ subset of the Baire space. A = p[T] for some tree T (constructible from r) on (ω, ω). (That is $x \in A$ iff from some y, $((x_0,y_0),(x_1,y_1),\ldots)$ is a path through T.)

Given a partial play s, let T_s be the subtree of T consistent with s subject to $\max(y_0,y_1,\ldots,y_{\text{len}(s)-1}) < \text{len}(s)$. The additional condition ensures that T_s is finite. Consistency means that every path through T_s is of the form $((x_0,y_0),(x_1,y_1),\ldots,(x_i,y_i))$ where $(x_0,x_1,\ldots,x_i)$ is an initial segment of s.

To prove that A is determined, define auxiliary game as follows:

In addition to ordinary moves, player 2 must play a mapping of T_s into ordinals (below a sufficiently large ordinal κ) such that:

- Each new move extends the previous mapping.
- The ordering of the ordinals agrees with the kleene–brouwer order on T_s.

Recall that Kleene–Brouwer order is like lexicographical order except that if s properly extends t then s < t. It is a well-ordering iff the tree is well-founded.

The auxiliary game is open. Proof: If player 2 does not lose at a finite stage, then the union of all T_s (which is the tree that corresponds to the play) is well-founded, and so the result of the non-auxiliary play is not in A.

Thus, the auxiliary game is determined. Proof: By transfinite induction, for each ordinal α compute the set of positions where player 1 can force a win in α steps, where a position with player 2 to move is losing (for player 2) in α steps iff for every move the resulting position is losing in less than α steps. One strategy for player 1 is to reduce α with each position (say picking the least α and breaking ties by picking the least move), and one strategy for player 2 is to pick the least (actually any would work) move that does not lead to a position with an α assigned. Note that L(r) contains the set of winning positions as well as the winning strategies given above.

A winning strategy for player 2 in the original game leads to winning strategy in the auxiliary game: The subtree of T corresponding to the winning strategy is well-founded, so player 2 can pick ordinals based on the Kleene–Brouwer order of the tree. Also, trivially, a winning strategy for player 2 in the auxiliary game gives a winning strategy for player 2 in original game.

It remains to show that using $r^{\#}$, the above-mentioned winning strategy for player 1 in the auxiliary game can be converted into a winning strategy in the original game. $r^{\#}$ gives a proper class I of (L(r),∈,r) indiscernible ordinals. By indiscernibility, if κ and the ordinals in the auxiliary response are in I, then the moves by player 1 do not depend on the auxiliary moves (or on κ), and so the strategy can be converted into a strategy for the original game (since player 2 can hold out with indiscernibles for any finite number of steps). Suppose that player 1 loses in the original game. Then, the tree corresponding to a play is well-founded. Therefore, player 2 can win the auxiliary game by using auxiliary moves based on the indiscernibles (since the order type of indiscernibles exceeds the Kleene–Brouwer order of the tree), which contradicts player 1 winning the auxiliary game.

Woodin Cardinals

If there is a Woodin cardinal with a measurable cardinal above it, then Π^1_2 determinacy holds. More generally, if there are n Woodin cardinals with a measurable cardinal above them all, then Π^1_{n+1} determinacy holds. From Π^1_{n+1} determinacy, it follows that there is a transitive inner model containing n Woodin cardinals.

Δ^1_2 (lightface) determinacy is equiconsistent with a Woodin cardinal. If Δ^1_2 determinacy holds, then for a Turing cone of x (that is for every real x of sufficiently high Turing degree), L[x] satisfies OD-determinacy (that is determinacy of games on integers of length ω and ordinal-definable payoff), and in $HOD^{L[x]}$ $\omega_2^{L[x]}$ is a Woodin cardinal.

Projective Determinacy

If there are infinitely many Woodin cardinals, then projective determinacy holds; that is, every game whose winning condition is a projective set is determined. From projective determinacy it follows that, for every natural number n, there is a transitive inner model that satisfies that there are n Woodin cardinals.

Axiom of Determinacy

The axiom of determinacy, or AD, asserts that every two-player game of perfect information of length ω, in which the players play naturals, is determined.

AD is provably false from ZFC; using the axiom of choice one may prove the existence of a non-determined game. However, if there are infinitely many Woodin cardinals with a measurable above them all, then L(R) is a model of ZF that satisfies AD.

Consequences of Determinacy

Regularity Properties for Sets of Reals

If A is a subset of Baire space such that the Banach–Mazur game for A is determined, then either II has a winning strategy, in which case A is meager, or I has a winning strategy, in which case A is comeager on some open neighborhood.

This does not quite imply that A has the property of Baire, but it comes close: A simple modification of the argument shows that if Γ is an adequate pointclass such that every game in Γ is determined, then every set of reals in Γ has the property of Baire.

In fact this result is not optimal; by considering the unfolded Banach–Mazur game we can show that determinacy of Γ (for Γ with sufficient closure properties) implies that every set of reals that is the projection of a set in Γ has the property of Baire. So for example the existence of a measurable cardinal implies Π^1_1 determinacy, which in turn implies that every Σ^1_2 set of reals has the property of Baire.

By considering other games, we can show that Π^1_n determinacy implies that every Σ^1_{n+1} set of reals has the property of Baire, is Lebesgue measurable (in fact universally measurable) and has the perfect set property.

Periodicity Theorems

The first periodicity theorem implies that, for every natural number n, if Δ^1_{2n+1} determinacy holds, then Π^1_{2n+1} and Σ^1_{2n+2} have the prewellordering property (and that Σ^1_{2n+1} and Π^1_{2n+2} do not have the prewellordering property, but rather have the separation property).

The second periodicity theorem implies that, for every natural number n, if Δ^1_{2n+1}

determinacy hold, then Π^1_{2n+1} and Σ^1_{2n} have the scale property. In particular, if projective determinacy holds, then every projective relation has a projective uniformization.

The third periodicity theorem gives a sufficient condition for a game to have a definable winning strategy.

Applications to Decidability of Certain Second-order Theories

In 1969, Michael O. Rabin proved that the second-order theory of n successors is decidable. A key component of the proof requires showing determinacy of parity games, which lie in the third level of the Borel hierarchy.

Wadge Determinacy

Wadge determinacy is the statement that for all pairs A, B of subsets of Baire space, the Wadge game G(A ,B) is determined. Similarly for a pointclass Γ, Γ Wadge determinacy is the statement that for all sets A, B in Γ, the Wadge game G(A, B) is determined.

Wadge determinacy implies the semilinear ordering principle for the Wadge order. Another consequence of Wadge determinacy is the perfect set property.

In general, Γ Wadge determinacy is a consequence of the determinacy of Boolean combinations of sets in Γ. In the projective hierarchy, Π^1_1 Wadge determinacy is equivalent to Π^1_1 determinacy, as proved by Leo Harrington. This result was extended by Hjorth to prove that Π^1_2 Wadge determinacy (and in fact the semilinear ordering principle for Π^1_2) already implies Π^1_2 determinacy.

Axiom of Projective Determinacy

In mathematical logic, projective determinacy is the special case of the axiom of determinacy applying only to projective sets.

The axiom of projective determinacy, abbreviated PD, states that for any two-player infinite game of perfect information of length ω in which the players play natural numbers, if the victory set (for either player, since the projective sets are closed under complementation) is projective, then one player or the other has a winning strategy.

The axiom is not a theorem of ZFC (assuming ZFC is consistent), but unlike the full axiom of determinacy (AD), which contradicts the axiom of choice, it is not known to be inconsistent with ZFC. PD follows from certain large cardinal axioms, such as the existence of infinitely many Woodin cardinals.

PD implies that all projective sets are Lebesgue measurable (in fact, universally measurable) and have the perfect set property and the property of Baire. It also implies that every projective binary relation may be uniformized by a projective set.

Axiom of Determinacy

In mathematics, the axiom of determinacy (abbreviated as AD) is a possible axiom for set theory introduced by Jan Mycielski and Hugo Steinhaus in 1962. It refers to certain two-person topological games of length ω. AD states that every game of a certain type is determined; that is, one of the two players has a winning strategy.

They motivated AD by its interesting consequences, and suggested that AD could be true in the least natural model L(R) of a set theory, which accepts only a weak form of the axiom of choice (AC) but contains all real and all ordinal numbers. Some consequences of AD followed from theorems proved earlier by Stefan Banach and Stanisław Mazur, and Morton Davis. Mycielski and Stanisław Świerczkowski contributed another one: AD implies that all sets of real numbers are Lebesgue measurable. Later Donald A. Martin and others proved more important consequences, especially in descriptive set theory. In 1988, John R. Steel and W. Hugh Woodin concluded a long line of research. Assuming the existence of some uncountable cardinal numbers analogous to $\aleph_0$, they proved the original conjecture of Mycielski and Steinhaus that AD is true in L(R).

Types of Game that is Determined

The axiom of determinacy refers to games of the following specific form: Consider a subset A of the Baire space ωω of all infinite sequences of natural numbers. Two players, I and II, alternately pick natural numbers:

$$n_0, n_1, n_2, n_3, \ldots$$

After infinitely many moves, a sequence $(n_i)_{i\in\omega}$ is generated. Player I wins the game if and only if the sequence generated is an element of A. The axiom of determinacy is the statement that all such games are determined.

Not all games require the axiom of determinacy to prove them determined. If the set A is clopen, the game is essentially a finite game, and is therefore determined. Similarly, if A is a closed set, then the game is determined. It was shown in 1975 by Donald A. Martin that games whose winning set is a Borel set are determined. It follows from the existence of sufficiently large cardinals that all games with winning set a projective set are determined, and that AD holds in L(R).

The axiom of determinacy implies that for every subspace X of the real numbers, the Banach–Mazur game BM(X) is determined (and therefore that every set of reals has the property of Baire).

Incompatibility of the Axiom of Determinacy with the Axiom of Choice

The set S1 of all first player strategies in an ω-game G has the same cardinality as the continuum. The same is true of the set S2 of all second player strategies. We note that the cardinality of the set SG of all sequences possible in G is also the continuum. Let A be the subset of SG of all sequences that make the first player win. With the axiom of choice we can well order the continuum; furthermore, we can do so in such a way that any proper initial portion does not have the cardinality of the continuum. We create a counterexample by transfinite induction on the set of strategies under this well ordering:

We start with the set A undefined. Let T be the "time" whose axis has length continuum. We need to consider all strategies {s1(T)} of the first player and all strategies {s2(T)} of the second player to make sure that for every strategy there is a strategy of the other player that wins against it. For every strategy of the player considered we will generate a sequence that gives the other player a win. Let t be the time whose axis has length $\aleph_0$ and which is used during each game sequence.

- Consider the current strategy {s1(T)} of the first player.
- Go through the entire game, generating (together with the first player's strategy s1(T)) a sequence {a(1), b(2), a(3), b(4),...,a(t), b(t+1),...}.
- Decide that this sequence does not belong to A, i.e. s1(T) lost.
- Consider the strategy {s2(T)} of the second player.
- Go through the next entire game, generating (together with the second player's strategy s2(T)) a sequence {c(1), d(2), c(3), d(4),...,c(t), d(t+1),...}, making sure that this sequence is different from {a(1), b(2), a(3), b(4),...,a(t), b(t+1),...}.
- Decide that this sequence belongs to A, i.e. s2(T) lost.
- Keep repeating with further strategies if there are any, making sure that sequences already considered do not become generated again. (We start from the set of all sequences and each time we generate a sequence and refute a strategy we project the generated sequence onto first player moves and onto second player moves, and we take away the two resulting sequences from our set of sequences).
- For all sequences that did not come up in the above consideration arbitrarily decide whether they belong to A, or to the complement of A.

Once this has been done we have a game G. If you give me a strategy s1 then we considered that strategy at some time T = T(s1). At time T, we decided an outcome of s1 that would be a loss of s1. Hence this strategy fails. But this is true for an arbitrary strategy; hence the axiom of determinacy and the axiom of choice are incompatible.

Infinite Logic and the Axiom of Determinacy

Many different versions of infinitary logic were proposed in the late 20th century. One reason that has been given for believing in the axiom of determinacy is that it can be written as follows (in a version of infinite logic):

$$\forall G \subseteq \mathrm{Seq}(S):$$

$$\forall a \in S : \exists a' \in S : \forall b \in S : \exists b' \in S : \forall c \in S : \exists c' \in S \ldots : (a, a', b, b', c, c' \ldots) \in G$$

Or

$$\exists a \in S : \forall a' \in S : \exists b \in S : \forall b' \in S : \exists c \in S : \forall c' \in S \ldots : (a, a', b, b', c, c' \ldots) \notin G$$

Seq(S) is the set of all ω-sequences of S. The sentences here are infinitely long with a countably infinite list of quantifiers where the ellipses appear.

In an infinitary logic, this principle is therefore a natural generalization of the usual (de Morgan) rule for quantifiers that are true for finite formulas, such as:

$$\forall a : \exists b : \forall c : \exists d : R(a, b, c, d)$$

Or

$$\exists a : \forall b : \exists c : \forall d : \neg R(a, b, c, d)$$

Large Cardinals and the Axiom of Determinacy

The consistency of the axiom of determinacy is closely related to the question of the consistency of large cardinal axioms. By a theorem of Woodin, the consistency of Zermelo–Fraenkel set theory without choice (ZF) together with the axiom of determinacy is equivalent to the consistency of Zermelo–Fraenkel set theory with choice (ZFC) together with the existence of infinitely many Woodin cardinals. Since Woodin cardinals are strongly inaccessible, if AD is consistent, then so are an infinity of inaccessible cardinals.

Moreover, if to the hypothesis of an infinite set of Woodin cardinals is added the existence of a measurable cardinal larger than all of them, a very strong theory of Lebesgue measurable sets of reals emerges, as it is then provable that the axiom of determinacy is true in L(R), and therefore that every set of real numbers in L(R) is determined.

Set-theoretic Topology

In mathematics, set-theoretic topology is a subject that combines set theory and general topology. It focuses on topological questions that are independent of Zermelo–Fraenkel set theory (ZFC).

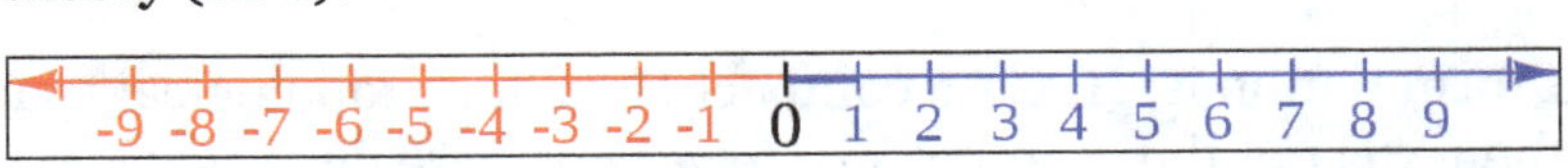

The space of integers has cardinality $\aleph_0$, while the real numbers has cardinality $2^{\aleph_0}$. The topologies of both spaces have cardinality $2^{\aleph_0}$. These are examples of cardinal functions, a topic in set-theoretic topology.

Objects Studied in Set-theoretic Topology

Dowker Spaces

In the mathematical field of general topology, a Dowker space is a topological space that is T4 but not countably paracompact.

Dowker conjectured that there were no Dowker spaces, and the conjecture was not resolved until M.E. Rudin constructed one in 1971. Rudin's counterexample is a very large space (of cardinality $\aleph_\omega^{\aleph_0}$ and is generally not well-behaved. Zoltán Balogh gave the first ZFC construction of a small (cardinality continuum) example, which was better-behaved than Rudin's. Using PCF theory, M. Kojman and S. Shelah constructed a subspace of Rudin's Dowker space of cardinality $\aleph_{\omega+1}$ that is also Dowker.

Normal Moore Spaces

A famous problem is the normal Moore space question, a question in general topology that was the subject of intense research. The answer to the normal Moore space question was eventually proved to be independent of ZFC.

Martin's Axiom

For any cardinal k, we define a statement, denoted by MA(k):

> "For any partial order P satisfying the countable chain condition (hereafter ccc) and any family D of dense sets in P such that $|D| \leq k$, there is a filter F on P such that $F \cap d$ is non-empty for every d in D."

Since it is a theorem of ZFC that MA(c) fails, the Martin's axiom is stated as:

Martin's axiom (MA): For every $k < c$, MA(k) holds.

In this case (for application of ccc), an antichain is a subset A of P such that any two distinct members of A are incompatible (two elements are said to be compatible if there exists a common element below both of them in the partial order). This differs from, for example, the notion of antichain in the context of trees.

MA($2^{\aleph_0}$) is false: [0, 1] is a compact Hausdorff space, which is separable and so ccc. It has no isolated points, so points in it are nowhere dense, but it is the union of $2^{\aleph_0}$ many points.

An equivalent formulation is: If X is a compact Hausdorff topological space which satisfies the ccc then X is not the union of k or fewer nowhere dense subsets.

Martin's axiom has a number of other interesting combinatorial, analytic and topological consequences:

- The union of k or fewer null sets in an atomless σ-finite Borel measure on a Polish space is null. In particular, the union of k or fewer subsets of R of Lebesgue measure 0 also has Lebesgue measure 0.
- A compact Hausdorff space X with $|X| < 2^k$ is sequentially compact, i.e., every sequence has a convergent subsequence.
- No non-principal ultrafilter on N has a base of cardinality < k.
- Equivalently for any x in βN\N we have $\chi(x) \geq k$, where χ is the character of x, and so $\chi(\beta N) \geq k$.
- MA($\aleph_1$) implies that a product of ccc topological spaces is ccc (this in turn implies there are no Suslin lines).
- MA + ¬CH implies that there exists a Whitehead group that is not free; Shelah used this to show that the Whitehead problem is independent of ZFC.

Forcing

In the mathematical discipline of set theory, forcing is a technique for proving consistency and independence results. It was first used by Paul Cohen in 1963, to prove the independence of the axiom of choice and the continuum hypothesis from Zermelo–Fraenkel set theory.

Forcing has been considerably reworked and simplified in the following years, and has since served as a powerful technique, both in set theory and in areas of mathematical logic such as recursion theory. Descriptive set theory uses the notions of forcing from both recursion theory and set theory. Forcing has also been used in model theory, but it is common in model theory to define genericity directly without mention of forcing.

Intuition

Intuitively, forcing consists of expanding the set theoretical universe V to a larger universe V*. In this bigger universe, for example, one might have many new subsets of $\omega = \{0,1,2,\ldots\}$ that were not there in the old universe, and thereby violate the continuum hypothesis.

While impossible when dealing with finite sets, this is just another version of Cantor's paradox about infinity. In principle, one could consider:

$$V^* = V \times \{0,1\},$$

Identify $x \in V$ with $(x,0)$, and then introduce an expanded membership relation involving "new" sets of the form $(x,1)$. Forcing is a more elaborate version of this idea, reducing the expansion to the existence of one new set, and allowing for fine control over the properties of the expanded universe.

Cohen's original technique, now called ramified forcing, is slightly different from the unramified forcing expounded here. Forcing is also equivalent to the method of Boolean-valued models, which some feel is conceptually more natural and intuitive, but usually much more difficult to apply.

Forcing Posets

A forcing poset is an ordered triple, $(\mathbb{P},\leq,1)$, where $\leq$ is a preorder on $\mathbb{P}$ that is atomless, meaning that it satisfies the following condition:

- For each $p\in\mathbb{P}$, there are $q,r\in\mathbb{P}$ such that $q,r\leq p$, with no $s\in\mathbb{P}$ such that $s\leq q,r$. The largest element of $\mathbb{P}$ is 1, that is, $p\leq 1$ for all $p\in\mathbb{P}$.

Members of $\mathbb{P}$ are called forcing conditions or just conditions. One reads $p\leq q$ as "p is stronger than q". Intuitively, the "smaller" condition provides "more" information, just as the smaller interval [3.1415926, 3.1415927] provides more information about the number π than the interval [3.1, 3.2] does.

There are various conventions in use. Some authors require $\leq$ to also being antisymmetric, so that the relation is a partial order. Some use the term partial order anyway, conflicting with standard terminology, while some use the term preorder. The largest element can be dispensed with. The reverse ordering is also used, most notably by Saharon Shelah and his co-authors.

P-names

Associated with a forcing poset $\mathbb{P}$ is the class $V^{(\mathbb{P})}$ of $\mathbb{P}$-names. A $\mathbb{P}$-name is a set A of the form:

$$A\subseteq\{(u,p)\mid u \text{ is a } \mathbb{P}\text{-name and } p\in\mathbb{P}\}$$

This is actually a definition by transfinite recursion. More precisely, one first uses transfinite recursion to define the following hierarchy:

$$\text{Name}(\varnothing)=\varnothing;$$

$$\text{Name}(\alpha+1)=\mathcal{P}(\text{Name}(\alpha)\times\mathbb{P}),\text{ where } \mathcal{P} \text{ denotes the power-set operator;}$$

$$\text{Name}(\lambda)=\bigcup\{\text{Name}(\alpha)\mid\alpha<\lambda\},\text{ if } \lambda \text{ is a limit ordinal.}$$

Then the class of $\mathbb{P}$-names is defined as:

$$V^{(\mathbb{P})}=\bigcup\{\text{Name}(\alpha)\mid\alpha \text{ is an ordinal}\}$$

The $\mathbb{P}$-names are, in fact, an expansion of the universe. Given $x \in V$, one defines $\check{x}$ to be the $\mathbb{P}$-name:

$$\check{x} = \{(\check{y},1) \mid y \in x\}$$

Again, this is really a definition by transfinite recursion.

Interpretation

Given any subset G of $\mathbb{P}$, one next defines the interpretation or valuation map from $\mathbb{P}$-names by:

$$\mathrm{val}(u,G) = \{\mathrm{val}(v,G) \mid \exists p \in G : (v,p) \in u\}$$

This is again a definition by transfinite recursion. Note that if $1 \in G$, then $\mathrm{val}(\check{x},G) = x$. One then defines:

$$\underline{G} = \{(\check{p},p) \mid p \in G\}$$

So that,

$$\mathrm{val}(\underline{G},G) = \{\mathrm{val}(\check{p},G) \mid p \in G\} = G.$$

Example:

A good example of a forcing poset is $(\mathrm{Bor}(I),\subseteq,I)$, where I = [0,1] and Bor(I) is the collection of Borel subsets of I having non-zero Lebesgue measure. In this case, one can talk about the conditions as being probabilities, and a Bor(I)-name assigns membership in a probabilistic sense. Due to the ready intuition this example can provide, probabilistic language is sometimes used with other divergent forcing posets.

Countable Transitive Models and Generic Filters

The key step in forcing is, given a ZFC universe V, to find an appropriate object G not in V. The resulting class of all interpretations of $\mathbb{P}$-names will be a model of ZFC that properly extends the original (since $G \notin V$).

Instead of working with V, it is useful to consider a countable transitive model M with $(\mathbb{P},\leq,1) \in M$. "Model" refers to a model of set theory, either of all of ZFC, or a model of a large but finite subset of ZFC, or some variant thereof. "Transitivity" means that if $x \in y \in M$, then $x \in M$. The Mostowski collapse lemma states that this can be assumed if the membership relation is well-founded. The effect of transitivity is that membership and other elementary notions can be handled intuitively. Countability of the model relies on the Löwenheim–Skolem theorem.

As M is a set, there are sets not in M – this follows from Russell's paradox. The appropriate set G to pick and adjoin to M is a generic filter on P. The "filter" condition means that:

- $G \subseteq \mathbb{P}$.
- $1 \in G$.
- If $p \geq q \in G$, then $p \in G$.
- If $p, q \in G$, then there exists an $r \in G$ such that $r \leq p, q$.

For G to be "generic" means:

- If $D \in M$ is a "dense" subset of $\mathbb{P}$ (that is, for each $p \in \mathbb{P}$, there exists a $q \in D$ such that $q \leq p$), then $G \cap D \neq \emptyset$.

The existence of a generic filter G follows from the Rasiowa–Sikorski lemma. In fact, slightly more is true: Given a condition $p \in \mathbb{P}$, one can find a generic filter G such that $p \in G$. Due to the splitting condition on $\mathbb{P}$ (termed being 'atomless'), if G is a filter, then $\mathbb{P} \backslash G$ is dense. If $G \in M$, then $\mathbb{P} \setminus G \in M$ because M is a model of ZFC. For this reason, a generic filter is never in M.

Consistency

The discussion above can be summarized by the fundamental consistency result that, given a forcing poset $\mathbb{P}$, we may assume the existence of a generic filter G, not belonging to the universe V, such that V[G] is again a set-theoretic universe that models ZFC. Furthermore, all truths in V[G] may be reduced to truths in V involving the forcing relation.

Both styles, adjoining G to either a countable transitive model M or the whole universe V, are commonly used. Less commonly seen is the approach using the "internal" definition of forcing, in which no mention of set or class models is made. This was Cohen's original method, and in one elaboration, it becomes the method of Boolean-valued analysis.

Cohen Forcing

The simplest nontrivial forcing poset is $(\mathrm{Fin}(\omega,2), \supseteq, 0)$, the finite partial functions from ω to $2 \stackrel{\mathrm{df}}{=} \{0,1\}$ under reverse inclusion. That is, a condition p is essentially two disjoint finite subsets $p^{-1}[1]$ and $p^{-1}[0]$ of ω, to be thought of as the "yes" and "no" parts of p, with no information provided on values outside the domain of p. "q is stronger than p" means that $q \supseteq p$, in other words, the "yes" and "no" parts of q are supersets of the "yes" and "no" parts of p, and in that sense, provide more information.

Let G be a generic filter for this poset. If p and q are both in G, then $p \cup q$ is a condition because G is a filter. This means that $g = \bigcup G$ is a well-defined partial function from ω to 2 because any two conditions in G agree on their common domain.

In fact, g is a total function. Given $n \in \omega$, let $D_n = \{p \mid p(n) \text{is defined}\}$. Then D_n is dense. (Given any p, if n is not in p's domain, adjoin a value for n—the result is in D_n. A condition $p \in G \cap D_n$ has n in its domain, and since $p \subseteq g$, we find that g(n) is defined.

Let, $X = g^{-1}[1]$, the set of all "yes" members of the generic conditions. It is possible to give a name for X directly. Let,

$$\underline{X} = \left\{ \left(\check{n}, p \right) \mid p(n) = 1 \right\}$$

Then, $\text{val}(\underline{X}, G) = X$. Now suppose that $A \subseteq \omega$ in V. We claim that $X \neq A$.

Let,

$$D_A = \{p \mid (\exists n)(n \in \text{Dom}(p) \wedge (p(n) = 1 \Leftrightarrow n \notin A))\}$$

Then D_A is dense. (Given any p, if n is not in its domain, adjoin a value for n contrary to the status of "$n \in A$".) Then any $p \in G \cap D_A$ witnesses $X \neq A$. To summarize, X is a "new" subset of ω, necessarily infinite.

Replacing ω with $\omega \times \omega_2$, that is, consider instead finite partial functions whose inputs are of the form (n, α), with $n < \omega$ and $\alpha < \omega_2$, and whose outputs are 0 or 1, one gets ω_2 new subsets of ω. They are all distinct, by a density argument: Given $\alpha < \beta < \omega_2$, let,

$$D_{\alpha,\beta} = \{p \mid (\exists n)(p(n, \alpha) \neq p(n, \beta))\},$$

Then each $D_{\alpha,\beta}$ is dense, and a generic condition in it proves that the αth new set disagrees somewhere with the βth new set.

This is not yet the falsification of the continuum hypothesis. One must prove that no new maps have been introduced which map ω onto ω_1, or ω_1 onto ω_2. For example, if one considers instead $\text{Fin}(\omega, \omega_1)$, finite partial functions from ω to ω_1, the first uncountable ordinal, one gets in V[G] a bijection from ω to ω_1. In other words, ω_1 has collapsed, and in the forcing extension, is a countable ordinal.

The last step in showing the independence of the continuum hypothesis, then, is to show that Cohen forcing does not collapse cardinals. For this, a sufficient combinatorial property is that all of the antichains of the forcing poset are countable.

Countable Chain Condition

An (strong) antichain A of $\mathbb{P}$ is a subset such that if $p, q \in A$, then p and q are incompatible (written $p \perp q$), meaning there is no r in $\mathbb{P}$ such that $r \leq p$ and $r \leq q$. In the example on Borel sets, incompatibility means that $p \cap q$ has zero measure. In the example on finite partial functions, incompatibility means that $p \cup q$ is not a function, in other words, p and q assign different values to some domain input.

$\mathbb{P}$ satisfies the countable chain condition (c.c.c.) if and only if every antichain in $\mathbb{P}$ is countable. (The name, which is obviously inappropriate, is a holdover from older terminology. Some mathematicians write "c.a.c." for "countable antichain condition".)

It is easy to see that Bor(I) satisfies the c.c.c. because the measures add up to at most 1. Also, Fin(E,2) satisfies the c.c.c., but the proof is more difficult.

Given an uncountable subfamily $W \subseteq \text{Fin}(E,2)$, shrink W to an uncountable subfamily W_0 of sets of size n, for some $n < \omega$. If $p(e_1) = b_1$ for uncountably many $p \in W_0$, shrink this to an uncountable subfamily W_1 and repeat, getting a finite set $\{(e_1, b_1),\ldots,(e_k, b_k)\}$ and an uncountable family W_k of incompatible conditions of size $n-k$ such that every e is in Dom(p) for at most countable many $p \in W_k$. Now, pick an arbitrary $p \in W_k$, and pick from W_k any q that is not one of the countably many members that have a domain member in common with p. Then $p \cup \{(e_1, b_1),\ldots,(e_k, b_k)\}$ and $q \cup \{(e_1, b_1),\ldots,(e_k, b_k)\}$ are compatible, so W is not an antichain. In other words, Fin(E,2)-antichains are countable.

The importance of antichains in forcing is that for most purposes, dense sets and maximal antichains are equivalent. A maximal antichain A is one that cannot be extended to a larger antichain. This means that every element $p \in \mathbb{P}$ is compatible with some member of A. The existence of a maximal antichain follows from Zorn's Lemma. Given a maximal antichain A, let:

$$D = \{p \in \mathbb{P} \mid (\exists q \in A)(p \le q)\}$$

Then D is dense, and $G \cap D \ne \varnothing$ if and only if $G \cap A \ne \varnothing$. Conversely, given a dense set D, Zorn's Lemma shows that there exists a maximal antichain $A \subseteq D$, and then $G \cap D \ne \varnothing$ if and only if $G \cap A \ne \varnothing$.

Assume that $\mathbb{P}$ satisfies the c.c.c. Given $x, y \in V$, with $f : x \to y$ a function in V[G], one can approximate f inside V as follows. Let u be a name for f (by the definition of V[G] and let p be a condition that forces u to be a function from x to y. Define a function F, whose domain is x, by:

$$F(a) \stackrel{df}{=} \left\{ b \mid (\exists q \in \mathbb{P}) \left[(q \le p) \wedge \left(q \Vdash u\left(\check{a}\right) = \check{b} \right) \right] \right\}$$

By the definability of forcing, this definition makes sense within V. By the coherence of forcing, a different b come from an incompatible p. By c.c.c., F(a) is countable.

In summary, f is unknown in V as it depends on G, but it is not wildly unknown for a c.c.c.-forcing. One can identify a countable set of guesses for what the value of f is at any input, independent of G.

This has the following very important consequence. If in V[G], $f:\alpha\to\beta$ is a surjection from one infinite ordinal onto another, then there is a surjection $g:\omega\times\alpha\to\beta$ in V, and consequently, a surjection $h:\alpha\to\beta$ in V. In particular, cardinals cannot collapse. The conclusion is that $2^{\aleph_0}\geq\aleph_2$ in V[G].

Easton Forcing

The exact value of the continuum in the above Cohen model, and variants like $\mathrm{Fin}(\omega\times\kappa,2)$ for cardinals κ in general, was worked out by Robert M. Solovay, who also worked out how to violate GCH (the generalized continuum hypothesis), for regular cardinals only, a finite number of times. For example, in the above Cohen model, if CH holds in V, then $2^{\aleph_0}=\aleph_2$ holds in V[G].

William B. Easton worked out the proper class version of violating the GCH for regular cardinals, basically showing that the known restrictions, (monotonicity, Cantor's Theorem and König's Theorem), were the only ZFC-provable restrictions.

Easton's work was notable in that it involved forcing with a proper class of conditions. In general, the method of forcing with a proper class of conditions fails to give a model of ZFC. For example, forcing with $\mathrm{Fin}(\omega\times\mathrm{On},2)$, where On is the proper class of all ordinals, makes the continuum a proper class. On the other hand, forcing with $\mathrm{Fin}(\omega,\mathrm{On})$ introduces a countable enumeration of the ordinals. In both cases, the resulting V[G] is visibly not a model of ZFC.

At one time, it was thought that more sophisticated forcing would also allow an arbitrary variation in the powers of singular cardinals. However, this has turned out to be a difficult, subtle and even surprising problem, with several more restrictions provable in ZFC and with the forcing models depending on the consistency of various large-cardinal properties. Many open problems remain.

Random Reals

Random forcing can be defined as forcing over the set P of all compact subsets of [0,1] of positive measure ordered by relation $\subseteq$ (smaller set in context of inclusion is smaller set in ordering and represents condition with more information). There are two types of important dense sets:

- For any positive integer n the set:

$$D_n=\left\{p\in P:\mathrm{diam}(p)<\frac{1}{n}\right\}$$

is dense, where (p) is diameter of the set p.

- For any Borel subset $B \subseteq [0,1]$ of measure 1, the set:

  $D_B = \{p \in P : p \subseteq B\}$

  is dense.

For any filter G and for any finitely many elements $p_1,...,p_n \in G$ there is $q \in G$ such that holds $q \le p_1,...,p_n$. In case of this ordering, this means that any filter is set of compact sets with finite intersection property. For this reason, intersection of all elements of any filter is nonempty. Let G is filter intersecting dense set D_n for any positive integer n, then filter G contains conditions of arbitrary small positive diameter. Therefore, intersection of all conditions from G has diameter 0. Only nonempty sets of diameter 0 are singletons. Finally, there is exactly one real number r such that $r \in \bigcap G$.

Let $B \subseteq [0,1]$ is any Borel set of measure 1. If G intersects D_B, then $r \in B$.

However, generic filter over countable transitive model V is not in V. Real r is defined by G. One can also prove that it is not in V. The problem is that if $p \in P$, then $V \models$ "p is compact", but from the viewpoint of universe p can be non-compact and the intersection of all conditions from generic filter is actually empty. For this reason, we consider set $C = \{\overline{p} : p \in G\}$ of closures (in topological sense) of conditions from generic filter. Due to $\overline{p} \supseteq p$ and finite intersection property of G, the set C also has finite intersection property. Elements of the set C are bounded closed sets as closures of bounded sets. Therefore, C is set of compacts with finite intersection property and for this reason has nonempty intersection. Due to $\text{diam}(\overline{p}) = \text{diam}(p)$ and model V inherits metric from universe, the set C has elements of arbitrary small diameter. Finally, there is exactly one real that belongs to all members of the set C. The generic filter G can be reconstructed from r as $G = \{p \in P : r \in \overline{p}\}$.

If a is name of r, and for $B \in V$ V holds $V \models$ "B is Borel set of measure 1", then holds:

$$V[G] \models \left(p \Vdash_{\mathbb{P}} a \in \check{B} \right)$$

or some $p \in G$. There is name a such that for any generic filter G holds:

$$\text{val}(a,G) \in \bigcup_{p \in G} \overline{p}$$

Then,

$$V[G] \models \left(p \Vdash_{\mathbb{P}} a \in \check{B} \right)$$

Holds for any condition p.

Every Borel set can, non-uniquely, be built up, starting from intervals with rational endpoints and applying the operations of complement and countable unions, a countable number of times. The record of such a construction is called a Borel code. Given a Borel set B in V, one recovers a Borel code, and then applies the same construction sequence in V[G], getting a Borel set B*. It can be proven that one gets the same set independent of the construction of B, and that basic properties are preserved. For example, if $B \subseteq C$, then $B^* \subseteq C^*$. If B has measure zero, then B* has measure zero. This mapping $B \mapsto B^*$ is injective.

For any set $B \subseteq [0,1]$ such that $B \in V$ and $V \models$ "B is Borel set of measure 1" holds $r \in B^*$.

This means that r is "infinite random sequence of 0s and 1s" from the viewpoint of V, which means that it satisfies all statistical tests from the ground model V.

So given r, a random real, one can show that:

$$G = \{B \text{ (in V)} \mid r \in B^* \text{ (in V[G])}\}$$

Because of the mutual inter-definability between r and G, one generally writes $V[r]$ and $V[G]$.

A different interpretation of reals in V[G] was provided by Dana Scott. Rational numbers in V[G] have names that correspond to countably-many distinct rational values assigned to a maximal antichain of Borel sets – in other words, a certain rational-valued function on I = [0,1]. Real numbers in V[G] then correspond to Dedekind cuts of such functions, that is, measurable functions.

Boolean-valued Models

Perhaps more clearly, the method can be explained in terms of Boolean-valued models. In these, any statement is assigned a truth value from some complete atomless Boolean algebra, rather than just a true/false value. Then an ultrafilter is picked in this Boolean algebra, which assigns values true/false to statements of our theory. The point is that the resulting theory has a model which contains this ultrafilter, which can be understood as a new model obtained by extending the old one with this ultrafilter. By picking a Boolean-valued model in an appropriate way, we can get a model that has the desired property. In it, only statements which must be true (are "forced" to be true) will be true, in a sense (since it has this extension/minimality property).

Meta-mathematical Explanation

In forcing, we usually seek to show that some sentence is consistent with ZFC (or optionally some extension of ZFC. One way to interpret the argument is to assume that ZFC is consistent and then prove that ZFC combined with the new sentence is also consistent.

Each "condition" is a finite piece of information – the idea is that only finite pieces are relevant for consistency, since, by the compactness theorem, a theory is satisfiable if and only if every finite subset of its axioms is satisfiable. Then we can pick an infinite set of consistent conditions to extend our model. Therefore, assuming the consistency of ZFC, we prove the consistency of ZFC extended by this infinite set.

Logical Explanation

By Gödel's second incompleteness theorem, one cannot prove the consistency of any sufficiently strong formal theory, such as ZFC, using only the axioms of the theory itself, unless the theory is inconsistent. Consequently, mathematicians do not attempt to prove the consistency of ZFC using only the axioms of ZFC, or to prove that ZFC + H is consistent for any hypothesis H using only ZFC + H. For this reason, the aim of a consistency proof is to prove the consistency of ZFC + H relative to the consistency of ZFC. Such problems are known as problems of relative consistency, one of which proves:

$$(*)\mathrm{ZFC} \vdash \mathrm{Con(ZFC)} \rightarrow \mathrm{Con(ZFC+H)}$$

The general schema of relative consistency proofs follows. As any proof is finite, it uses only a finite number of axioms:

$$\mathrm{ZFC} + \neg\mathrm{Con(ZFC+H)} \vdash (\exists \mathrm{T})(\mathrm{Fin(T)} \wedge \mathrm{T} \subseteq \mathrm{ZFC} \wedge (\mathrm{T} \vdash \neg \mathrm{H}))$$

For any given proof, ZFC can verify the validity of this proof. This is provable by induction on the length of the proof:

$$\mathrm{ZFC} \vdash (\forall \mathrm{T})((\mathrm{T} \vdash \neg \mathrm{H}) \rightarrow (\mathrm{ZFC} \vdash (\mathrm{T} \vdash \neg \mathrm{H})))$$

Then resolve:

$$\mathrm{ZFC} + \neg\mathrm{Con(ZFC+H)} \vdash (\exists \mathrm{T})(\mathrm{Fin(T)} \wedge \mathrm{T} \subseteq \mathrm{ZFC} \wedge (\mathrm{ZFC} \vdash (\mathrm{T} \vdash \neg \mathrm{H})))$$

By proving the following:

$$(**)\ \mathrm{ZFC} \vdash (\forall \mathrm{T})(\mathrm{Fin(T)} \wedge \mathrm{T} \subseteq \mathrm{ZFC} \rightarrow (\mathrm{ZFC} \vdash \mathrm{Con(T+H)}))$$

It can be concluded that:

$$\mathrm{ZFC} + \neg\mathrm{Con(ZFC+H)} \vdash (\exists \mathrm{T})(\mathrm{Fin(T)} \wedge \mathrm{T} \subseteq \mathrm{ZFC} \wedge (\mathrm{ZFC} \vdash (\mathrm{T} \vdash \neg \mathrm{H})) \wedge (\mathrm{ZFC} \vdash \mathrm{Con(T+H)}))$$

which is equivalent to:

$$\mathrm{ZFC} + \neg\mathrm{Con(ZFC+H)} \vdash \neg\mathrm{Con(ZFC)}$$

which gives (*). The core of the relative consistency proof is proving (**). A ZFC proof of Con(T+H) can be constructed for any given finite subset T of the ZFC axioms (by ZFC instruments of course). (No universal proof of Con(T+H) of course).

In ZFC, it is provable that for any condition p, the set of formulas (evaluated by names) forced by p is deductively closed. Furthermore, for any ZFC axiom, ZFC proves that this axiom is forced by 1. Then it suffices to prove that there is at least one condition that forces H.

In the case of Boolean-valued forcing, the procedure is similar: proving that the Boolean value of H is not 0.

Another approach uses the Reflection Theorem. For any given finite set of ZFC axioms, there is a ZFC proof that this set of axioms has a countable transitive model. For any given finite set T of ZFC axioms, there is a finite set T' of ZFC axioms such that ZFC proves that if a countable transitive model M satisfies T', then M[G] satisfies T. By proving that there is finite set T'' of ZFC axioms such that if a countable transitive model M satisfies T'', then M[G] satisfies the hypothesis H. Then, for any given finite set T of ZFC axioms, ZFC proves Con(T+H) Con(T+H).

Sometimes in (**), a stronger theory S than ZFC is used for proving Con(T+H). Then we have proof of the consistency of ZFC + H relative to the consistency of S. Note that $ZFC \vdash Con(ZFC) \leftrightarrow Con(ZFL)$, where ZFL is ZF + V = L (the axiom of constructibility).

General Set Theory

General set theory (GST) is George Boolos's name for a fragment of the axiomatic set theory Z. GST is sufficient for all mathematics not requiring infinite sets, and is the weakest known set theory whose theorems include the Peano axioms.

Ontology

The ontology of GST is identical to that of ZFC, and hence is thoroughly canonical. GST features a single primitive ontological notion, that of set, and a single ontological assumption, namely that all individuals in the universe of discourse (hence all mathematical objects) are sets. There is a single primitive binary relation, set membership; that set a is a member of set b is written $a \in b$ (usually read "a is an element of b").

Axioms

The symbolic axioms below are from Boolos, and govern how sets behave and interact.

The natural language versions of the axioms are intended to aid the intuition. The background logic is first order logic with identity.

- Axiom of Extensionality: The sets x and y are the same set if they have the same members.

  $$\forall x \forall y [\forall z [z \in x \leftrightarrow z \in y] \rightarrow x = y]$$

  The converse of this axiom follows from the substitution property of equality.

- Axiom Schema of Specification (or Separation or Restricted Comprehension): If z is a set and φ is any property which may be satisfied by all, some, or no elements of z, then there exists a subset y of z containing just those elements x in z which satisfy the property φ. The restriction to z is necessary to avoid Russell's paradox and its variants. More formally, let φ(x) be any formula in the language of GST in which x may occur freely and y does not. Then all instances of the following schema are axioms:

  $$\forall z \exists y \forall x [x \in y \leftrightarrow (x \in z \wedge \phi(x))]$$

- Axiom of Adjunction: If x and y are sets, then there exists a set w, the adjunction of x and y, whose members are just y and the members of x.

  $$\forall x \forall y \exists w \forall z [z \in w \leftrightarrow (z \in x \vee z = y)]$$

  Adjunction refers to an elementary operation on two sets, and has no bearing on the use of that term elsewhere in mathematics, including in category theory.

GST is the fragment of Z obtained by omitting the axioms Union, Power Set, Elementary Sets (essentially Pairing) and Infinity, then taking Adjunction, a theorem of Z, as an axiom. The result is a first order theory.

Setting φ(x) in Separation to x ≠ x, and assuming that the domain is nonempty, assures the existence of the empty set. Adjunction implies that if x is a set, then so is $S(x) = x \cup \{x\}$. Given Adjunction, the usual construction of the successor ordinals from the empty set can proceed, one in which the natural numbers are defined as $\varnothing, S(\varnothing), S(S(\varnothing)), \ldots$, etc. More generally, given any model M of ZFC, the collection of hereditarily finite sets in M will satisfy the GST axioms. Therefore, GST cannot prove the existence of even a countable infinite set, that is, of a set whose cardinality is $\aleph_0$. Even if GST did afford a countably infinite set, GST could not prove the existence of a set whose cardinality is $\aleph_1$, because GST lacks the axiom of power set. Hence GST cannot ground analysis and geometry, and is too weak to serve as a foundation for mathematics.

Boolos was interested in GST only as a fragment of Z that is just powerful enough to interpret Peano arithmetic. He never lingered over GST, only mentioning it briefly in several papers discussing the systems of Frege's Grundlagen and Grundgesetze, and how they could be modified to eliminate Russell's paradox. The system $A\xi'[\delta_0]$ in Tarski and Givant is essentially GST with an axiom schema of induction replacing Specification, and with the existence of an empty set explicitly assumed.

GST is called STZ in Burgess, Burgess's theory ST is GST with Empty Set replacing the axiom schema of specification. That the letters "ST" also appear in "GST" is a coincidence.

Metamathematics

The most remarkable fact about ST (and hence GST), is that these tiny fragments of set theory give rise to such rich metamathematics. While ST is a small fragment of the well-known canonical set theories ZFC and NBG, ST interprets Robinson arithmetic (Q), so that ST inherits the nontrivial metamathematics of Q. For example, ST is essentially undecidable because Q is, and every consistent theory whose theorems include the ST axioms is also essentially undecidable. This includes GST and every axiomatic set theory worth thinking about, assuming these are consistent. In fact, the undecidability of ST implies the undecidability of first-order logic with a single binary predicate letter.

Q is also incomplete in the sense of Gödel's incompleteness theorem. Any axiomatizable theory, such as ST and GST, whose theorems include the Q axioms is likewise incomplete. Moreover, the consistency of GST cannot be proved within GST itself, unless GST is in fact inconsistent.

GST is:

- Mutually interpretable with Peano arithmetic (thus it has the same proof-theoretic strength as PA).
- Immune to the three great antinomies of naïve set theory: Russell's, Burali-Forti's, and Cantor's.
- Not finitely axiomatizable. Montague showed that ZFC is not finitely axiomatizable, and his argument carries over to GST. Hence any axiomatization of GST must either include at least one axiom schema such as Separation.
- Interpretable in relation algebra because no part of any GST axiom lies in the scope of more than three quantifiers. This is the necessary and sufficient condition given in Tarski and Givant.

Kripke–Platek Set Theory

The Kripke–Platek set theory (KP) is an axiomatic set theory developed by Saul Kripke and Richard Platek.

KP is considerably weaker than Zermelo–Fraenkel set theory (ZFC), and can be thought of as roughly the predicative part of ZFC. The consistency strength of KP with an axiom of infinity is given by the Bachmann–Howard ordinal. Unlike ZFC, KP does not include the power set axiom, and KP includes only limited forms of the axiom of separation and axiom of replacement from ZFC. These restrictions on the axioms of KP lead to close connections between KP, generalized recursion theory, and the theory of admissible ordinals.

Axioms of KP

- Axiom of extensionality: Two sets are the same if and only if they have the same elements.
- Axiom of induction: φ(a) being a formula, if for all sets x the assumption that φ(y) holds for all elements y of x entails that φ(x) holds, then φ(x) holds for all sets x.
- Axiom of empty set: There exists a set with no members, called the empty set and denoted {}. (Note: the existence of a member in the universe of discourse, i.e., $\exists x(x = x)$, is implied in certain formulations of first-order logic, in which case the axiom of empty set follows from the axiom of Σ_0-separation, and is thus redundant.)
- Axiom of pairing: If x, y are sets, then so is {x, y}, a set containing x and y as its only elements.
- Axiom of union: For any set x, there is a set y such that the elements of y are precisely the elements of the elements of x.
- Axiom of Σ_0-separation: Given any set and any Σo-formula φ(x), there is a subset of the original set containing precisely those elements x for which φ(x) holds. (This is an axiom schema.)
- Axiom of Σ_0-collection: Given any Σ_0-formula φ(x, y), if for every set x there exists a unique set y such that φ(x, y) holds, then for all sets u there exists a set v such that for every x in u there is a y in v such that φ(x, y) holds.

  Here, a Σ_0, or Π_0, or Δ_0 formula is one all of whose quantifiers are bounded. This means any quantification is the form $\forall u \in v$ or $\exists u \in v$. (More generally, we would say that a formula is Σ_{n+1} when it is obtained by adding existential quantifiers in front of a Π_n formula, and that it is Πn+1 when it is obtained by adding universal quantifiers in front of a Σ_n formula: this is related to the arithmetical hierarchy but in the context of set theory.)
- Some but not all authors include an axiom of infinity (in which case the empty set axiom is unnecessary).

These axioms are weaker than ZFC as they exclude the axioms of powerset, choice, and sometimes infinity. Also the axioms of separation and collection here are weaker than the corresponding axioms in ZFC because the formulas φ used in these are limited to bounded quantifiers only.

The axiom of induction in KP is stronger than the usual axiom of regularity (which amounts to applying induction to the complement of a set (the class of all sets not in the given set)).

Proof that Cartesian Products Exist

Theorem: If A and B are sets, then there is a set A×B which consists of all ordered pairs (a, b) of elements a of A and b of B.

Proof: The set {a} (which is the same as {a, a} by the axiom of extensionality) and the set {a, b} both exist by the axiom of pairing. Thus,

$$(a,b) := \{\{a\},\{a,b\}\}$$

Exists by the axiom of pairing as well.

A possible Δ_0 formula expressing that p stands for (a, b) is:

$$\exists r \in p(a \in r \wedge \forall x \in r(x = a)) \wedge \exists s \in p(a \in s \wedge b \in s \wedge \forall x \in s(x = a \vee x = b)) \wedge \\ \forall t \in p((a \in t \wedge \forall x \in t(x = a)) \vee (a \in t \wedge b \in t \wedge \forall x \in t(x = a \vee x = b)))$$

Thus a superset of A×{b} = {(a, b) |a in A} exists by the axiom of collection.

Denote the formula for p above by $\psi(a,b,p)$. Then the following formula is also Δ_0:

$$\exists a \in A \psi(a,b,p)$$

Thus A × {b} itself exists by the axiom of separation.

If v is intended to stand for A×{b}, then a Δ_0 formula expressing that is:

$$\forall a \in A \exists p \in v \psi(a,b,p) \wedge \forall p \in v \exists a \in A \psi(a,b,p)$$

Thus a superset of {A×{b} | b in B} exists by the axiom of collection.

Putting $\exists b \in B$ in B in front of that last formula and we get from the axiom of separation that the set {A×{b} |b in B} itself exists.

Finally, A×B = {A × {b} | b in B} exists by the axiom of union.

Admissible Sets

A set A, is called admissible if it is transitive and $\langle A, \in \rangle$ is a model of Kripke–Platek set theory.

An ordinal number α is called an admissible ordinal if Lα is an admissible set.

The ordinal α is an admissible ordinal if and only if α is a limit ordinal and there does not exist a $\gamma < \alpha$ for which there is a $\Sigma_1(L_\alpha)$ mapping from γ onto α. If M is a standard model of KP, then the set of ordinals in M is an admissible ordinal.

If L_α is a standard model of KP set theory without the axiom of Σ_0-collection, then it is said to be an "amenable set".

Venn Diagram

A Venn diagram is a diagrammatic representation of ALL the possible relationships between different sets of a finite number of elements. Venn diagrams were conceived around 1880 by John Venn, an English logician, and philosopher. They are extensively used to teach Set Theory. A Venn diagram is also known as a Primary diagram, Set diagram or Logic diagram.

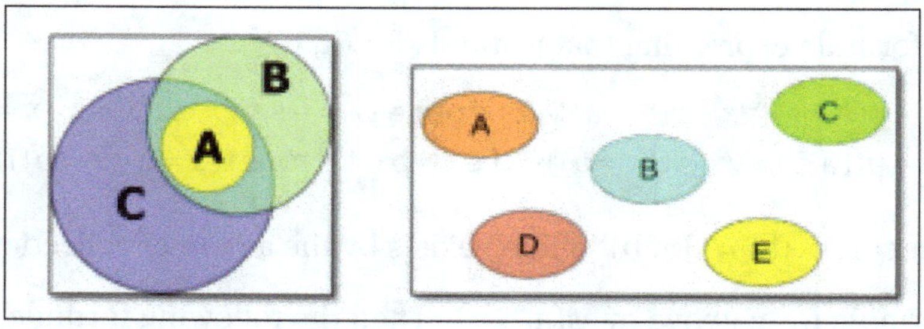

Representation of Sets in a Venn Diagram

It is done as per the following:

- Each individual set is represented mostly by a circle and enclosed within a quadrilateral (the quadrilateral represents the finiteness of the Venn diagram as well as the Universal set.)
- Labelling is done for each set with the set's name to indicate difference and the respective constituting elements of each set are written within the circles.
- Sets having no element in common are represented separately while those having some of the elements common within them are shown with overlapping.
- The elements are written within the circle representing the set containing them and the common elements are written in the parts of circles that are overlapped.

Operations on Venn Diagrams

Just like the mathematical operations on sets like Union, Difference, Intersection, Complement, etc. we have operations on Venn diagrams that are given as follows:

Union of Sets

Let A = {2, 4, 6, 8} and B = {6, 8, 10, 12}. Represent A U B through a well-labeled Venn diagram.

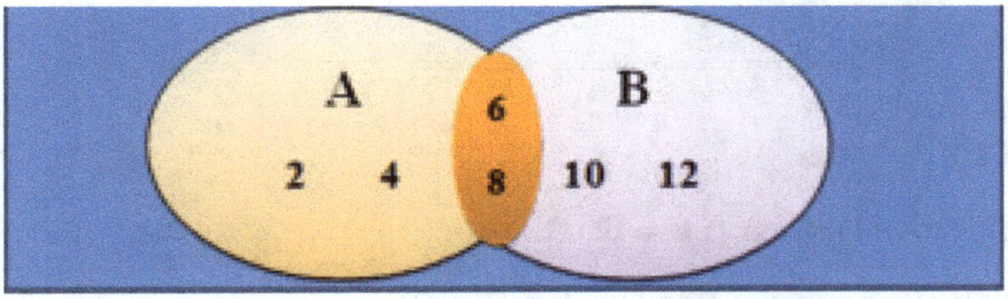

The orange colored patch represents the common elements {6, 8} and the quadrilateral represents A U B.

Properties of A U B

- The commutative law holds true as A U B = B U A.
- The associative law also holds true as (A U B) U C = A U (B U C).
- A U φ = A (Law of identity element).
- Idempotent Law – A U A = A.
- Law of the Universal Set U – A U U = U.

Intersection of Sets

An intersection is nothing but the collection of all the elements that are common to all the sets under consideration. Let A = {2, 4, 6, 8} and B = {6, 8, 10, 12} then A ∩ B is represented through a Venn diagram as per following:

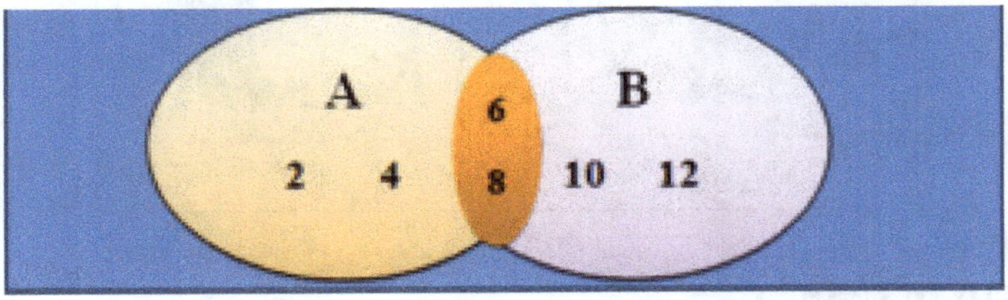

The orange colored patch represents the common elements {6, 8} as well as the A ∩ B.

The intersection of 2 or more sets is the overlapped parts of the individual circles with the elements written in the overlapped parts. Example:

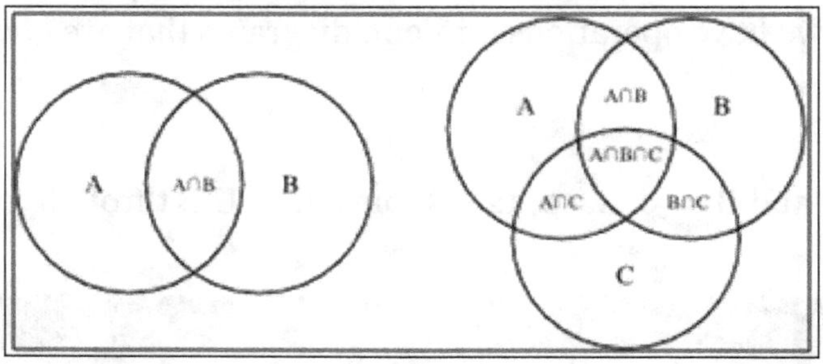

Properties of $A \cap B$

- Commutative law – $A \cap B = B \cap A$.
- Associative law – $(A \cap B) \cap C = A \cap (B \cap C)$.
- $\varphi \cap A = \varphi$.
- $U \cap A = A$.
- $A \cap A = A$; Idempotent law.
- Distributive law – $A \cap (B \cap C) = (A \cap B) \cup (A \cap C)$.

Difference of Sets

The difference of set A and B is represented as: $A - B = \{x: x \in A \text{ and } x \in B\}$ {converse holds true for $B - A$}. Let, $A = \{1, 2, 3, 4, 5, 6\}$ and $B = \{2, 4, 6, 8\}$ then $A - B = \{1, 3, 5\}$ and $B - A = \{8\}$. The sets $(A - B)$, $(B - A)$ and $(A \cap B)$ are mutually disjoint sets.

It means that there is NO element common to any of the three sets and the intersection of any of the two or all the three sets will result in a null or void or empty set. $A - B$ and $B - A$ are represented through Venn diagrams as follows:

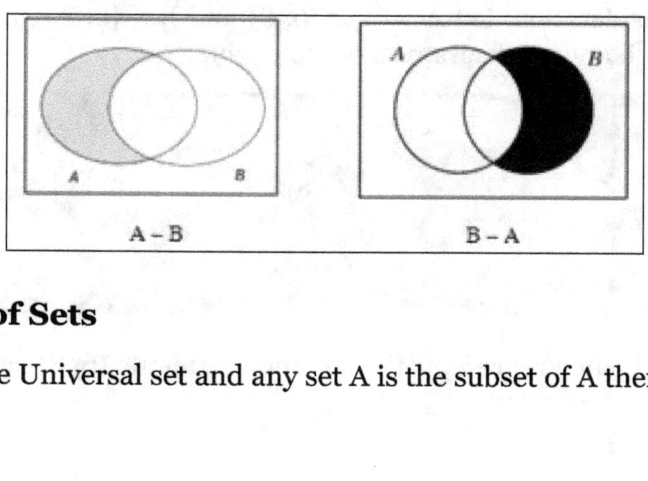

Complement of Sets

If U represents the Universal set and any set A is the subset of A then the complement of set A (represented as A') will contain ALL the elements which belong to the Universal set U but NOT to set A.

Mathematically – A' = U – A

Alternatively, the complement of a set A, A' is the difference between the universal set U and the set A. Example: Let universal set U = {1, 2, 3, 4, 5, 6, 7, 8, 9, 10} and set A = {1, 3, 5, 7, 9}, then complement of A is given as: A' = U – A = {2, 4, 6, 8, 10}.

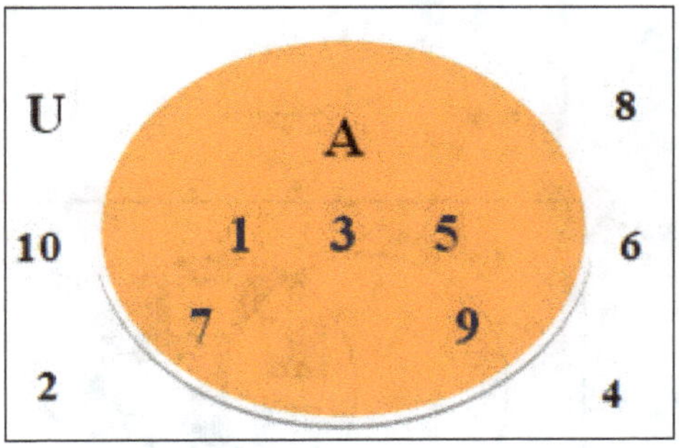

Properties of Complement Sets

- A U A' = U.
- A ∩ A' = φ.
- De Morgan's Law – (A U B)' = A' ∩ B' OR (A ∩ B)' = A' U B'.
- Law of double complementation: (A')' = A.
- φ' = U.
- U' = φ.

Example: Represent the Universal Set (U) = {x : x is an outcome of a dice's roll} and set A = {s : s ϵ Even numbers} through a Venn diagram.

Solution: Since, U = {1, 2, 3, 4, 5, 6} and A = {2, 4, 6}. Representing this with a Venn diagram we have:

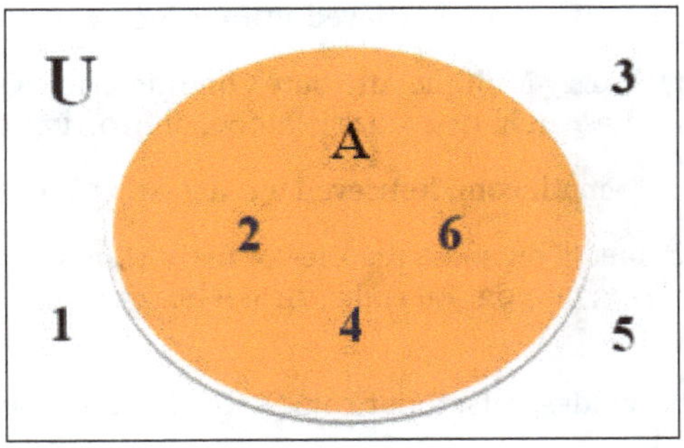

Representing them in Venn diagrams:

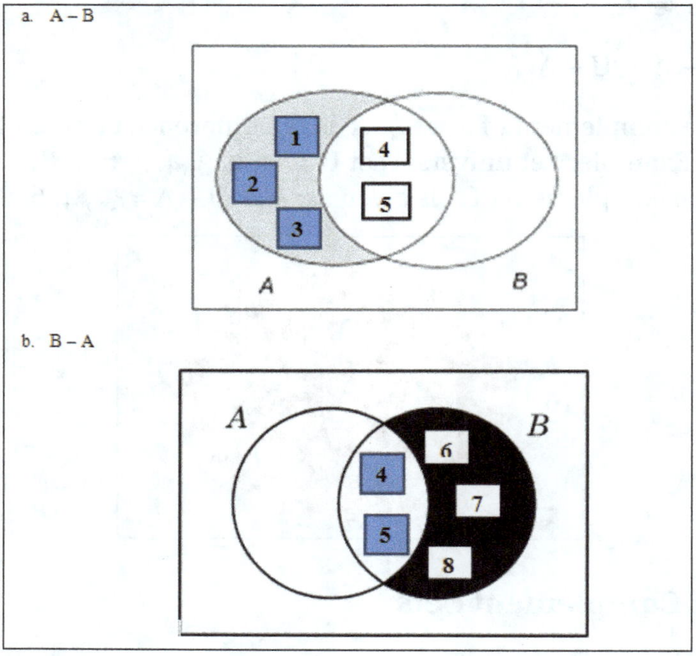

Here, A is a subset of U, represented as – A ⊂ U or,

U is the superset of A, represented as – U⊃ A.

If A = {1, 2, 3, 4, 5} and B = {4, 5, 6, 7, 8},

then represent A – B and B – A through Venn diagrams.

A – B = {1, 2, 3}

B – A = {6, 7, 8}

References

- Siegfried Gottwald, 2001. A Treatise on Many-Valued Logics. Baldock, Hertfordshire, England: Research Studies Press Ltd., ISBN 978-0-86380-262-1
- Set-theory, science: britannica.com, Retrieved 17 June, 2020
- Veri, Francesco (2017). "Fuzzy Multiple Attribute Conditions in fsQCA: Problems and Solutions". Sociological Methods & Research: 004912411772969. doi:10.1177/0049124117729693
- Types-of-sets: math-only-math.com, Retrieved 23 August, 2020
- Gerla, G. (2016). "Comments on some theories of fuzzy computation". International Journal of General Systems. 45 (4): 372–392. Bibcode:2016IJGS...45..372G. doi:10.1080/03081079.2015.1076403
- Venn-diagrams, maths, guides, sets: toppr.com, Retrieved 20 February, 2020

3

Model Theory

The study of classes of groups, fields, graphs and other aspects of set theory under mathematical logic is termed as model theory. Finite model theory, first-order logic, Gödel's completeness and incompleteness theorem, etc. are studied within it. The topics elaborated in this chapter will help in gaining a better perspective of model theory.

An interpretation of a formal language satisfying certain axioms. The basic formal language is the first-order language L_Ω of a given signature Ω including predicate symbols R_i, $i \in I$, function symbols f_j, $j \in J$ and constants c_k, $k \in K$. A model of the language L_Ω is an algebraic system of signature Ω.

Let Σ be a set of closed formulas in L_Ω. A model for Σ is a model for L_Ω in which all formulas from Σ are true. A set Σ is called consistent if it has at least one model. The class of all models of Σ is denoted by Mod Σ. Consistency of a set Σ means that Mod $\Sigma \neq \emptyset$.

A class K of models of a language L_Ω is called axiomatizable if there is a set Σ of closed formulas of L_Ω such that $K = Mod\Sigma$. The set $T(K)$ of all closed formulas of L_Ω that are true in each model of a given class K of models of L_Ω is called the elementary theory of K. Thus, a class K of models of L_Ω is axiomatizable if and only if $K = Mod\ T(K)$. If a class K consists of models isomorphic to a given model, then its elementary theory is called the elementary theory of this model.

Let A be a model of L_Ω having universe A. One may associate to each element $\alpha \in A$ a constant c_α and consider the first-order language $L_{\Omega A}$ of signature ΩA which is obtained from Ω by adding the constants c_α, $\alpha \in A$. $L_{\Omega A}$ is called the diagram language of the model A. The set $O(A)$ of all closed formulas of $L_{\Omega A}$ which are true in A on replacing each constant c_α by the corresponding element $\alpha \in A$ is called the description (or elementary diagram) of A. The set $D(A)$ of those formulas from $O(A)$ which are atomic or negations of atomic formulas is called the diagram of A.

Along with models of first-order languages, models of other types (infinitary logic, intuitionistic logic, many-sorted logic, second-order logic, many-valued logic, and modal logic) have also been considered.

The origins of model theory go back to the 1920's and 1930's, when the following two fundamental theorems were proved.

- Gödel compactness theorem: If each finite subcollection of a collection T of propositions in a first-order language is consistent, then the whole collection T is consistent.
- Löwenheim–Skolem theorem: If a collection of propositions in a first-order language of signature Ω has an infinite model, then it has a model of any infinite cardinality not less than the cardinality of Ω.

Gödel compactness theorem has had extensive application in algebra. On the basis of this theorem, A.I. Mal'tsev created a method of proof of local theorems in algebra.

Let A be an algebraic system of signature Ω, let |A| be the underlying set of A, let X ⊆ |A|, let (Ω, X) denote the signature obtained from Ω by the addition of symbols for distinguished elements c_a for all a ∈ X, and let (A, X) denote the algebraic system of signature (Ω, X) which is an enrichment of A in which for each a ∈ X the symbol c_a is interpreted by the element a. The set O(A) of all closed formulas of the signature (Ω,|A|) in a first-order language which are true in the system (A, |A|) is called the elementary diagram of the algebraic system A (or the description of the algebraic system A), and the set D(A) of those formulas from O(A) which are either atomic or the negation of an atomic formula is called the diagram of A. An algebraic system B is called an elementary extension of A if |A| ⊆ |B| and if (B,|A|) is a model for O(A). In this case A is called an elementary subsystem of B. For example, the set of rational numbers with the usual order relation is an elementary subsystem of the system of real numbers with the usual order relation.

A subsystem A of an algebraic system B of signature Ω is an elementary subsystem of B if and only if for each closed formula $(\exists v)\Phi(v)$ in the first-order language of signature (Ω, |A|) which is true in (B,|A|) there is an a ∈|A| such that Φ (c_a) is true in (B, |A|). It follows at once from this criterion that the union of an increasing chain of elementary subsystems is an elementary extension of each of these subsystems. If a closed ∀∃-formula in a first-order language is true in every system of an increasing chain of systems, then it is true in the union of the chain.

Let the signature Ω contain a one-place relation symbol *U*. One says that a model *A* of a theory *T* of signature Ω has type (α, β) if the cardinality of |*A*| is equal to α and if the cardinality of $U(A) = \{a \in |A| : A \models U(a)\}$ is equal to β. Vaught's theorem: If an elementary theory T of countable signature has a model of type (α, β) where α > β, then *T* has a model of type $(\aleph_1, \aleph_0)$. Under the assumption that the generalized continuum hypothesis holds, an elementary theory of countable signature has models of types $(\aleph_{\alpha+2},$

$\aleph_{\alpha+1}$) for each α, if it has a model of type ($\aleph_1$, $\aleph_0$). Under the same assumption the theory Th(*A*), where the signature of *A* is (+, ., 0, 1, <, *U*), |*A*| is the set of all real numbers, *U(A)* the set of all integers, and +, ., 0, 1, <, are defined in the usual way, does not have a model of type ($\aleph_2$, $\aleph_0$).

Let (*A, P*) denote the enrichment of the algebraic system *A* by a predicate *P*, and let Ω, *P*) be the signature obtained from Ω by the addition of the predicate symbol *P*. In many cases it is important to understand when in each member of a class K of algebraic systems of signature (Ω, *P*) the predicate *P* is given by a formula in the first-order language of the signature Ω. A partial answer to this question is given by Beth's definability theorem: There exists a formula $\Phi(x)$ in the first-order language of signature Ω such that the formula $(\forall x)\ (\Phi(x) \leftrightarrow P(x))$ is true for all members of an axiomatizable class K of signature (Ω, P) if and only if the set $\{P{:}(A, P) \in K\}$ contains at most one element for each algebraic system A of signature Ω.

Much research in model theory is connected with the study of properties preserved under operations on algebraic systems. The most important operations include homomorphism, direct and filtered products.

A statement Φ is said to be stable with respect to homomorphisms if the truth of Φ in an algebraic system A implies the truth of Φ in all epimorphic images of *A*. A formula Φ in a first-order language is called positive if Φ does not contain negation and implication signs. It has been proved that a statement Φ in a first-order language is stable relative to homomorphisms if and only if Φ is equivalent to a positive statement. A similar theorem holds for the language $L_{\omega_1\omega}$.

A formula $\Phi(x_1,...,x_n)$ in a first-order language of signature Ω is called a Horn formula it it can be obtained by conjunction and quantification from formulas of the form $(\Phi_1 \& ... \& \Phi_s) \rightarrow \Phi$, $\neg(\Phi_1 \& ... \& \Phi_s)$, where $\Phi_1,...,\Phi_s$ are atomic formulas in the first-order language of Ω. Examples of Horn formulas are identities and quasi-identities. Central in the theory of ultraproducts is the theorem of J. Łos: Every formula in a first-order language is stable with respect to any ultrafilter. A formula in a first-order language is conditionally stable with respect to any filter if and only if it is equivalent to a Horn formula. There is the following theorem: Two algebraic systems *A* and *B* of signature Ω are elementarily equivalent if and only if there is an ultrafilter *D* on a set I such that A^I/D and B^I/D are isomorphic. The cardinality of a filtered product is countably infinite if for each natural number *n* the number of factors of cardinality *n* is finite. If for each natural number *n* the set of indices for which the corresponding factors have cardinality *n* does not belong to D, then the cardinality of the ultraproduct with respect to a non-principal ultrafilter D on a countable set I is equal to that of the continuum. For each infinite set I of cardinality α there is a filter D on I such that for each filter D_1 on I containing D, and each infinite set A, the cardinality of A^I/D is not less than 2^α.

Many applications have been found for the Ehrenfeucht–Mostowski theorem on the existence of models with a large number of automorphisms: For any totally ordered set

X in an axiomatizable class K of algebraic systems containing an infinite system, there is a system A such that X ⊆ |A| and such that each order-preserving one-to-one mapping of X onto X can be extended to an automorphism of A.

The major notions in model theory are those of universal, homogeneous and saturated systems. Let A and B be algebraic systems of a signature Ω. A mapping *f* of a set X ⊆ |A| into a set Y ⊆ |B| is called elementary if for each formula $\Phi(x_1,\ldots,x_n)$ in the first-order language of the signature Ω and any $a_1,\ldots,a_n \in X$ the equivalence $A \models \Phi(x_1,\ldots,x_n) \Leftrightarrow B \models \Phi(f(a_1),\ldots,f(a_n))$ holds. A system A is called α-universal if for every system B that is elementarily equivalent to A and of cardinality not exceeding α, there is an elementary mapping from |B| into |A|. A system A is called α-homogeneous if for every set X ⊆ |A| of cardinality less than α, every elementary mapping from X into |A| can be extended to an elementary mapping of |A| onto |A| (that is, to an automorphism of A). A system A of signature Ω is called α-saturated if for every set X ⊆ |A| of cardinality less than α and every collection Σ of formulas in the first-order language of the signature (Ω, X) not containing free variables other than x_0, Σ finitely satisfiable in (A, X) implies that Σ is satisfiable in (A, X). A system A is called universal (respectively, homogeneous or saturated) if A is α-universal (respectively, α-homogeneous or α-saturated), where α is the cardinality of |A|. A system is saturated if and only if it is simultaneously universal and homogeneous. Two elementary equivalent saturated systems of the same cardinality are isomorphic. All uncountable models of elementary theories which are categorical in uncountable cardinalities and of countable signature are saturated. A large number of examples of α-saturated systems is given by ultraproducts. For example, if D is a non-principal ultrafilter on a countable set *I*, then $\prod_{i\in I} A_i / D$ is an $\aleph_1$-saturated system for any algebraic system $A_i (i \in I)$ of a countable signature Ω.

The basic problems of model theory are the study of the expressive possibilities of a formalized language and the study of classes of structures defined by means of such a language.

The basic apparatus for the study of stable theories is the classification of formulas and locally consistent sets of formulas in these theories. Such a classification can be obtained by means of ascribing to formulas their ranks. Such ranks are usually ordinals and the ranking functions are given with the help of special topologies and other means. The study of ranking functions and their improvements is a rich source of information on the theories.

In the study of classes of models one is concerned with the number of distinct models, up to isomorphism, of a theory of a given cardinality, the existence of special models, for example, simple, minimal, saturated, homogeneous, universal, etc., and one creates means for constructing them.

The classical examples of application of methods of model theory are the papers of A. Robinson and his school, which developed an independent science — non-standard analysis; from the work of Mal'tsev and his school the applications of model-theoretic methods to topological algebra have been developed; the latest results on the properties of stable theories have been used in the study of concrete algebraic questions.

The above problems arose also in the study of various non-elementary languages, for example, obtained by the addition of new quantifiers, the introduction of infinite expressions, modalities, etc.

Finite Model Theory

Finite model theory (FMT) is a subarea of model theory (MT). MT is the branch of mathematical logic which deals with the relation between a formal language (syntax) and its interpretations (semantics). FMT is a restriction of MT to interpretations on finite structures, which have a finite universe.

- Since many central theorems of MT do not hold when restricted to finite structures, FMT is quite different from MT in its methods of proof. Central results of classical model theory that fail for finite structures under FMT include the compactness theorem, Gödel's completeness theorem, and the method of ultraproducts for first-order logic (FO).
- As MT is closely related to mathematical algebra, FMT became an "unusually effective" instrument in computer science. In other words: "In the history of mathematical logic most interest has concentrated on infinite structures....Yet, the objects computers have and hold are always finite. To study computation we need a theory of finite structures." Thus the main application areas of FMT are: descriptive complexity theory, database theory and formal language theory.
- FMT is mainly about discrimination of structures. The usual motivating question is whether a given class of structures can be described (up to isomorphism) in a given language. For instance, can all cyclic graphs be discriminated (from the non-cyclic ones) by a sentence of the first-order logic of graphs? This can also be phrased as: is the property "cyclic" FO expressible?

Basic Challenges

A single finite structure can always be axiomatized in first-order logic, where axiomatized in a language L means described uniquely up to isomorphism by a single L-sentence. Similarly, any finite collection of finite structures can always be axiomatized in first-order logic. Some, but not all, infinite collections of finite structures can also be axiomatized by a single first-order sentence.

Characterisation of a Single Structure

Is a language L expressive enough to axiomatize a single finite structure S?

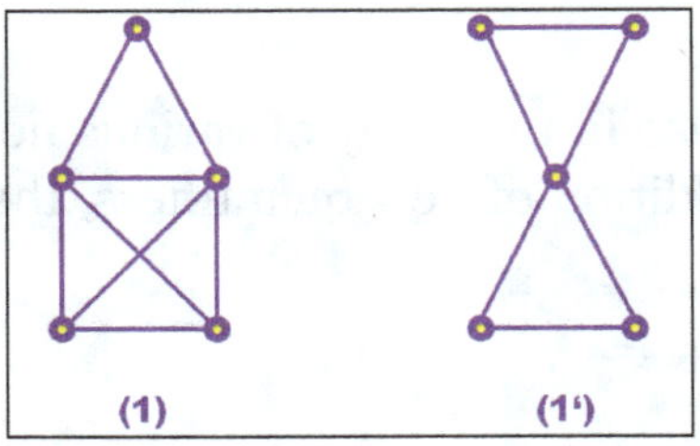

Single graphs (1) and (1') having common properties.

Problem: A structure like (1) in the figure can be described by FO sentences in the logic of graphs like:

- Every node has an edge to another node: $\forall_x \exists_y G(x,y)$.
- No node has an edge to itself: $\forall_{x,y} (G(x,y) \Rightarrow x \neq y)$.
- There is at least one node that is connected to all others: $\exists_x \forall_y (x \neq y \Rightarrow G(x,y))$.

However, these properties do not axiomatize the structure, since for structure (1') the above properties hold as well, yet structures (1) and (1') are not isomorphic.

Informally the question is whether by adding enough properties, these properties together describe exactly (1) and are valid (all together) for no other structure (up to isomorphism).

Approach: For a single finite structure it is always possible to precisely describe the structure by a single FO sentence. The principle is illustrated here for a structure with one binary relation R and without constants:

- Say that there are at least n elements: $\varphi 1 = \underset{i \neq j}{\wedge} \neg (x_i = x_j)$.
- Say that there are at most n elements: $\varphi 2 = \forall_y \underset{i}{\vee} (x_i = y)$.
- State every element of the relation R: $\varphi 3 = \underset{(a_i, a_j) \in R}{\wedge} R(x_i, x_j)$.
- State every non-element of the relation R: $\varphi 4 = \underset{(a_i, a_j) \notin R}{\wedge} \neg R(x_i, x_j)$.

All for the same tuple $x_1 .. x_n$, yielding the FO sentence $\exists x_1 .. \exists x_n (\varphi 1 \wedge \varphi 2 \wedge \varphi 3 \wedge \varphi 4)$.

Extension to a Fixed Number of Structures

The method of describing a single structure by means of a first-order sentence can easily be extended for any fixed number of structures. A unique description can be obtained by the disjunction of the descriptions for each structure. For instance, for two structures A and B with defining sentences φ_A and φ_B this would be $\varphi_A \vee \varphi_B$.

Extension to an Infinite Structure

By definition, a set containing an infinite structure falls outside the area that FMT deals with. Note that infinite structures can never be discriminated in FO, because of the Löwenheim–Skolem theorem, which implies that no first-order theory with an infinite model can have a unique model up to isomorphism.

The most famous example is probably Skolem's theorem, that there is a countable non-standard model of arithmetic.

Characterisation of a Class of Structures

Is a language L expressive enough to describe exactly (up to isomorphism) those finite structures that have certain property P?

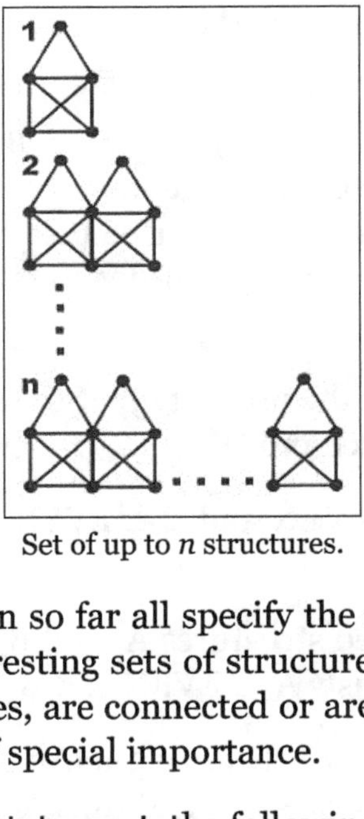

Set of up to n structures.

Problem: The descriptions given so far all specify the number of elements of the universe. Unfortunately most interesting sets of structures are not restricted to a certain size, like all graphs that are trees, are connected or are acyclic. Thus to discriminate a finite number of structures is of special importance.

Approach: Instead of a general statement, the following is a sketch of a methodology to differentiate between structures that can and cannot be discriminated.

- The core idea is that whenever one wants to see if a property P can be expressed in FO, one chooses structures A and B, where A does have P and B doesn't. If for A and B the same FO sentences hold, then P cannot be expressed in FO. In short:

  $A \in P$, $B \notin P$ and $A \equiv B$,

where $A \equiv B$ is shorthand for $A \models \alpha \Leftrightarrow B \models \alpha$ for all FO-sentences α, and P represents the class of structures with property P.

- The methodology considers countably many subsets of the language, the union of which forms the language itself. For instance, for FO consider classes FO[m] for each m. For each m the above core idea then has to be shown. That is:

  $A \in P$, $B \notin P$ and $A \equiv_m B$

  with a pair A, B for each m and α (in ≡) from FO[m]. It may be appropriate to choose the classes FO[m] to form a partition of the language.

- One common way to define FO[m] is by means of the quantifier rank qr(α) of a FO formula α, which expresses the depth of quantifier nesting. For example, for a formula in prenex normal form, qr is simply the total number of its quantifiers. Then FO[m] can be defined as all FO formulas α with qr(α) ≤ m (or, if a partition is desired, as those FO formulas with quantifier rank equal to m).

- Thus it all comes down to showing $A \models \alpha \Leftrightarrow B \models \alpha$ on the subsets FO[m]. The main approach here is to use the algebraic characterization provided by Ehrenfeucht–Fraïssé games. Informally, these take a single partial isomorphism on A and B and extend it m times, in order to either prove or disprove $A \equiv_m B$, dependent on who wins the game.

Example: We want to show that the property that the size of an ordered structure A = (A, ≤) is even, cannot be expressed in FO.

- The idea is to pick A ∈ EVEN and B ∉ EVEN, where EVEN is the class of all structures of even size.

- We start with two ordered structures A_2 and B_2 with universes A_2 = {1, 2, 3, 4} and B_2 = {1, 2, 3}. Obviously A_2 ∈ EVEN and B_2 ∉ EVEN.

- For m = 2, we can now show* that in a 2-move Ehrenfeucht–Fraïssé game on A_2 and B_2 the duplicator always wins, and thus A_2 and B_2 cannot be discriminated in FO[2], i.e. $A_2 \models \alpha \Leftrightarrow B_2 \models \alpha$ for every α ∈ FO[2].

- Next we have to scale the structures up by increasing m. For example, for m = 3 we must find an A_3 and B_3 such that the duplicator always wins the 3-move game. This can be achieved by A_3 = {1, ..., 8} and B_3 = {1, ..., 7}. More generally, we can choose A_m = {1, ..., 2^m} and B_m = {1, ..., 2^m-1}; for any m the duplicator always wins the m-move game for this pair of structures.

- Thus EVEN on finite ordered structures cannot be expressed in FO.

Applications

Database Theory

A substantial fragment of SQL (namely that which is effectively relational algebra) is based on first-order logic (more precisely can be translated in domain relational calculus by means of Codd›s theorem), as the following example illustrates: Think of a database table "GIRLS" with the columns "FIRST_NAME" and "LAST_NAME". This corresponds to a binary relation, say G(f, l) on FIRST_NAME X LAST_NAME. The FO query {l : G('Judy', l)}, which returns all the last names where the first name is 'Judy', would look in SQL like this:

```
select LAST_NAME

from GIRLS

where FIRST_NAME = 'Judy'
```

We assume here, that all last names appear only once (or we should use SELECT DISTINCT since we assume that relations and answers are sets, not bags).

Next we want to make a more complex statement. Therefore, in addition to the "GIRLS" table we have a table "BOYS" also with the columns "FIRST_NAME" and "LAST_NAME". Now we want to query the last names of all the girls that have the same last name as at least one of the boys. The FO query is $\{(f,l) : \exists h\ (G(f, l) \wedge B(h, l))\}$, and the corresponding SQL statement is:

```
select FIRST_NAME, LAST_NAME

from GIRLS

where LAST_NAME IN ( select LAST_NAME from BOYS );
```

Notice that in order to express the "∧" we introduced the new language element "IN" with a subsequent select statement. This makes the language more expressive for the price of higher difficulty to learn and implement. This is a common trade-off in formal language design. The way shown above ("IN") is by far not the only one to extend the language. An alternative way is e.g. to introduce a "JOIN" operator, that is:

```
select distinct g.FIRST_NAME, g.LAST_NAME

from GIRLS g, BOYS b

where g.LAST_NAME=b.LAST_NAME;
```

First-order logic is too restrictive for some database applications, for instance because of its inability to express transitive closure. This has led to more powerful constructs being added to database query languages, such as recursive WITH in SQL:1999. More expressive logics, like fixpoint logics, have therefore been studied in finite model theory because of their relevance to database theory and applications.

Querying and Search

Narrative data contains no defined relations. Thus the logical structure of text search queries can be expressed in propositional logic, like in:

```
("Java" AND NOT "island") OR ("C#" AND NOT "music")
```

Interpretation Model Theory

In model theory, interpretation of a structure *M* in another structure *N* (typically of a different signature) is a technical notion that approximates the idea of representing *M* inside *N*. For example every reduct or definitional expansion of a structure *N* has an interpretation in *N*.

Many model-theoretic properties are preserved under interpretability. For example if the theory of *N* is stable and *M* is interpretable in *N*, then the theory of *M* is also stable.

An interpretation of *M* in *N* with parameters (or without parameters, respectively) is a pair (n, f) where n is a natural number and f is a surjective map from a subset of N^n onto *M* such that the f-preimage (more precisely the f^k-preimage) of every set $X \subseteq M^k$ definable in *M* by a first-order formula without parameters is definable (in *N*) by a first-order formula with parameters (or without parameters, respectively). Since the value of n for an interpretation (n, f) is often clear from context, the map f itself is also called an interpretation.

To verify that the preimage of every definable (without parameters) set in *M* is definable in *N* (with or without parameters), it is sufficient to check the preimages of the following definable sets:

- The domain of *M*.
- The diagonal of *M*.
- Every relation in the signature of *M*.
- The graph of every function in the signature of *M*.

In model theory the term *definable* often refers to definability with parameters; if this convention is used, definability without parameters is expressed by the term *0-definable*. Similarly, an interpretation with parameters may be referred to as simply an interpretation, and an interpretation without parameters as a 0-interpretation.

Bi-interpretability

If *L, M* and *N* are three structures, *L* is interpreted in *M,* and *M* is interpreted in *N,* then one can naturally construct a composite interpretation of L in N. If two structures M and N are interpreted in each other, then by combining the interpretations in two possible ways, one obtains an interpretation of each of the two structures in itself. This observation permits one to define an equivalence relation among structures, reminiscent of the homotopy equivalence among topological spaces.

Two structures M and N are bi-interpretable if there exists an interpretation of M in N and an interpretation of N in M such that the composite interpretations of M in itself and of N in itself are definable in M and in N, respectively (the composite interpretations being viewed as operations on M and on N).

Example: The partial map f from Z × Z onto Q which maps (x, y) to x/y provides an interpretation of the field Q of rational numbers in the ring Z of integers (to be precise, the interpretation is $(2, f)$). In fact, this particular interpretation is often used to *define* the rational numbers. To see that it is an interpretation (without parameters), one needs to check the following preimages of definable sets in Q:

- The preimage of Q is defined by the formula $\varphi(x, y)$ given by $\neg\ (y = 0)$.
- The preimage of the diagonal of Q is defined by the formula $\varphi(x_1, y_1, x_2, y_2)$ given by $x_1 \times y_2 = x_2 \times y_1$.
- The preimages of 0 and 1 are defined by the formulas $\varphi(x, y)$ given by $x = 0$ and $x = y$.
- The preimage of the graph of addition is defined by the formula $\varphi(x_1, y_1, x_2, y_2, x_3, y_3)$ given by $x_1{\times}y_2{\times}y_3 + x_2{\times}y_1{\times}y_3 = x_3{\times}y_1{\times}y_2$.
- The preimage of the graph of multiplication is defined by the formula $\varphi(x_1, y_1, x_2, y_2, x_3, y_3)$ given by $x_1{\times}x_2{\times}y_3 = x_3{\times}y_1{\times}y_2$.

Reduct

In universal algebra and in model theory, a reduct of an algebraic structure is obtained by omitting some of the operations and relations of that structure. The converse of "reduct" is "expansion."

Let A be an algebraic structure (in the sense of universal algebra) or equivalently a structure in the sense of model theory, organized as a set X together with an indexed family of operations and relations φ_i on that set, with index set I. Then the reduct of A defined by a subset J of I is the structure consisting of the set X and J-indexed family of operations and relations whose j-th operation or relation for $j \in J$ is the j-th operation or relation of A. That is, this reduct is the structure A with the omission of those operations and relations φ_i for which i is not in J.

A structure A is an expansion of B just when B is a reduct of A. That is, reduct and expansion are mutual converses.

Example: The monoid (Z, +, 0) of integers under addition is a reduct of the group (Z, +, −, 0) of integers under addition and negation, obtained by omitting negation. By contrast, the monoid (N, +, 0) of natural numbers under addition is not the reduct of any group.

Conversely the group (Z, +, −, 0) is the expansion of the monoid (Z, +, 0), expanding it with the operation of negation.

Type of Model Theory

In model theory and related areas of mathematics, a type is an object that, loosely speaking, describes how a (real or possible) element or finite collection of elements in a mathematical structure might behave. More precisely, it is a set of first-order formulas in a language L with free variables $x_1, x_2, \ldots, x_n$ that are true of a sequence of elements of an L-structure $\mathcal{M}$. Depending on the context, types can be complete or partial and they may use a fixed set of constants, A, from the structure $\mathcal{M}$. The question of which types represent actual elements of $\mathcal{M}$ leads to the ideas of saturated models and omitting types.

Consider a structure $\mathcal{M}$ for a language L. Let M be the universe of the structure. For every $A \subseteq M$, let $L(A)$ be the language obtained from L by adding a constant c_a for every $a \in A$. In other words,

$$L(A) = L \cup \{c_a : a \in A\}.$$

A 1-type (of $\mathcal{M}$) over A is a set $p(x)$ of formulas in $L(A)$ with at most one free variable x (therefore 1-type) such that for every finite subset $p_0(x) \subseteq p(x)$ there is some $b \in M$, depending on $p_0(x)$, with $\mathcal{M} \vDash p_0(b)$ (i.e. all formulas in $p_0(x)$ are true in $\mathcal{M}$ when x is replaced by b).

Similarly an n-type (of $\mathcal{M}$) over A is defined to be a set $p(x_1, \ldots, x_n) = p(x)$ of formulas in $L(A)$, each having its free variables occurring only among the given n free variables $x_1, \ldots, x_n$, such that for every finite subset $p_0(x) \subseteq p(x)$ there are some elements $b_1, \ldots, b_n \in M$ with $\mathcal{M} \vDash p_0(b_1, \ldots, b_n)$.

A complete type of $\mathcal{M}$ over A is one that is maximal with respect to inclusion. Equivalently, for every $\phi(x) \in L(A, x)$ either $\phi(x) \in p(x)$ or $\neg\phi(x) \in p(x)$. Any non-complete type is called a partial type. So, the word type in general refers to any n-type, partial or complete, over any chosen set of parameters (possibly the empty set).

An n-type $p(x)$ is said to be realized in $\mathcal{M}$ if there is an element $b \in M^n$ such that $\mathcal{M} \models p(b)$. The existence of such a realization is guaranteed for any type by the compactness theorem, although the realization might take place in some elementary extension of $\mathcal{M}$, rather than in $\mathcal{M}$ itself. If a complete type is realized by b in $\mathcal{M}$, then the type is typically denoted $tp_n^{\mathcal{M}}(b/A)$ and referred to as the complete type of b over A.

Atypep(x)issaidtobeisolatedby φ,for $\varphi \in p(x)$, if $\forall \psi(x) \in p(x)$, $\mathrm{Th}(\mathcal{M}) \models \varphi(x) \rightarrow \psi(x)$. Since finite subsets of a type are always realized in $\mathcal{M}$, there is always an element $b \in M^n$ such that $\varphi(b)$ is true in $\mathcal{M}$; i.e. $\mathcal{M} \models \varphi(b)$, thus b realizes the entire isolated type. So isolated types will be realized in every elementary substructure or extension. Because of this, isolated types can never be omitted.

A model that realizes the maximum possible variety of types is called a saturated model, and the ultrapower construction provides one way of producing saturated models.

Examples of Types

Consider the language with one binary connective, which we denote as $\in$. Let $\mathcal{M}$ be the structure $\langle w, \in_w \rangle$ for this language, which is the ordinal w with its standard well-ordering. Let $\mathcal{T}$ denote the theory of $\mathcal{M}$.

Consider the set of formulas $p(x) := \{n \in_w x \mid n \in w\}$. First, we claim this is a type. Let $p_0(x) \subseteq p(x)$ be a finite subset of $p(x)$. We need to find a $b \in \omega$ that satisfies all the formulas in p_0. Well, we can just take the successor of the largest ordinal mentioned in the set of formulas $p_0(x)$. Then this will clearly contain all the ordinals mentioned in $p_0(x)$. Thus we have that $p(x)$ is a type. Next, note that $p(x)$ is not realized in $\mathcal{M}$. For, if it were there would be some $n \in w$ that contains every element of w. If we wanted to realize the type, we might be tempted to consider the model $\langle \omega+1, \in_{\omega+1} \rangle$, which is indeed a supermodel of $\mathcal{M}$ that realizes the type. Unfortunately, this extension is not elementary, that is, this model does not have to satisfy $\mathcal{T}$. In particular, the sentence $\exists x \forall y\ (y \in x \vee y = x)$ is satisfied by this model and not by $\mathcal{M}$.

So, we wish to realize the type in an elementary extension. We can do this by defining a new structure in the language, which we will denote $\mathcal{M}'$. The domain of the structure will be $\omega \cup \mathbb{Z}'$ where $\mathbb{Z}'$ is the set of integers adorned in such a way that $\mathbb{Z}' \cap \omega = \varnothing$. Let $<$ denote the usual order of $\mathbb{Z}'$. We interpret the symbol $\in$ in our new structure by $\in_{\mathcal{M}'} = \in_\omega \cup < \cup (\omega \times \mathbb{Z}')$. The idea being that we are adding a "$\mathbb{Z}$-chain", or copy of the integers, above all the finite ordinals. Clearly any element of $\mathbb{Z}'$ realizes the type $p(x)$. Moreover, one can verify that this extension is elementary.

Another example: the complete type of the number 2 over the empty set, considered as a member of the natural numbers, would be the set of all first-order statements, describing a variable x, that are true when $x = 2$. This set would include formulas such as $x \neq 1+1+1$, $x \leq 1+1+1+1+1$, and $\exists y(y < x)$. This is an example of an isolated type, since, working over the theory of the naturals, the formula $x = 1+1$ implies all other formulas that are true about the number 2.

As a further example, the statements,

$$\forall y(y^2 < 2 \Rightarrow y < x)$$

and,

$$\forall y((y > 0 \wedge y^2 > 2) \Rightarrow y > x)$$

describing the square root of 2 are consistent with the axioms of ordered fields, and can be extended to a complete type. This type is not realized in the ordered field of rational numbers, but is realized in the ordered field of reals. Similarly, the infinite set of formulas (over the empty set) {x>1, x>1+1, x>1+1+1, ...} is not realized in the ordered field of real numbers, but is realized in the ordered field of hyperreals. If we allow parameters, for instance all of the reals, we can specify a type $\{0 < x < r : r \in \mathbb{R}\}$ that is realized by an infinitesimal hyperreal that violates the Archimedean property.

The reason it is useful to restrict the parameters to a certain subset of the model is that it helps to distinguish the types that can be satisfied from those that cannot. For example, using the entire set of real numbers as parameters one could generate an uncountably infinite set of formulas like $x \neq 1$, $x \neq \pi$, ... that would explicitly rule out every possible real value for *x*, and therefore could never be realized within the real numbers.

Stone Spaces

It is useful to consider the set of complete *n*-types over *A* as a topological space. Consider the following equivalence relation on formulas in the free variables $x_1, \ldots, x_n$ with parameters in *A*:

$$\psi \equiv \phi \Leftrightarrow \mathcal{M} \models \forall x_1, \ldots, x_n(\psi(x_1, \ldots, x_n) \leftrightarrow \phi(x_1, \ldots, x_n)).$$

One can show that $\psi \equiv \phi$ if and only if they are contained in exactly the same complete types.

The set of formulas in free variables $x_1, \ldots, x_n$ over *A* up to this equivalence relation is a Boolean algebra (and is canonically isomorphic to the set of *A*-definable subsets of M^n). The complete *n*-types correspond to ultrafilters of this Boolean algebra. The set of complete *n*-types can be made into a topological space by taking the sets of types containing a given formula as basic open sets. This constructs the Stone space, which is compact, Hausdorff, and totally disconnected.

Example: The complete theory of algebraically closed fields of characteristic 0 has quantifier elimination, which allows one to show that the possible complete 1-types (over the empty set) correspond to:

- Roots of a given irreducible non-constant polynomial over the rationals with leading coefficient 1. For example, the type of square roots of 2. Each of these types is an open point of the Stone space.
- Transcendental elements, that are not roots of any non-zero polynomial. This type is a point in the Stone space that is closed but not open.

In other words, the 1-types correspond exactly to the prime ideals of the polynomial ring Q[x] over the rationals Q: if r is an element of the model of type p, then the ideal corresponding to p is the set of polynomials with r as a root (which is only the zero polynomial if r is transcendental). More generally, the complete n-types correspond to the prime ideals of the polynomial ring Q[$x_1, \ldots, x_n$], in other words to the points of the prime spectrum of this ring. (The Stone space topology can in fact be viewed as the Zariski topology of a Boolean ring induced in a natural way from the Boolean algebra. While the Zariski topology is not in general Hausdorff, it is in the case of Boolean rings). For example, if $q(x, y)$ is an irreducible polynomial in two variables, there is a 2-type whose realizations are (informally) pairs (x, y) of elements with $q(x, y)=0$.

Omitting Types Theorem

Given a complete n-type p one can ask if there is a model of the theory that omits p, in other words there is no n-tuple in the model that realizes p. If p is an isolated point in the Stone space, i.e. if $\{p\}$ is an open set, it is easy to see that every model realizes p (at least if the theory is complete). The omitting types theorem says that conversely if p is not isolated then there is a countable model omitting p (provided that the language is countable).

Example: In the theory of algebraically closed fields of characteristic 0, there is a 1-type represented by elements that are transcendental over the prime field. This is a non-isolated point of the Stone space (in fact, the only non-isolated point). The field of algebraic numbers is a model omitting this type, and the algebraic closure of any transcendental extension of the rationals is a model realizing this type.

All the other types are "algebraic numbers" (more precisely, they are the sets of first-order statements satisfied by some given algebraic number), and all such types are realized in all algebraically closed fields of characteristic 0.

Gödel's Completeness Theorem

Gödel's completeness theorem is a fundamental theorem in mathematical logic that establishes a correspondence between semantic truth and syntactic provability in first-order logic. It makes a close link between model theory that deals with what is true in different models, and proof theory that studies what can be formally proven in particular formal systems.

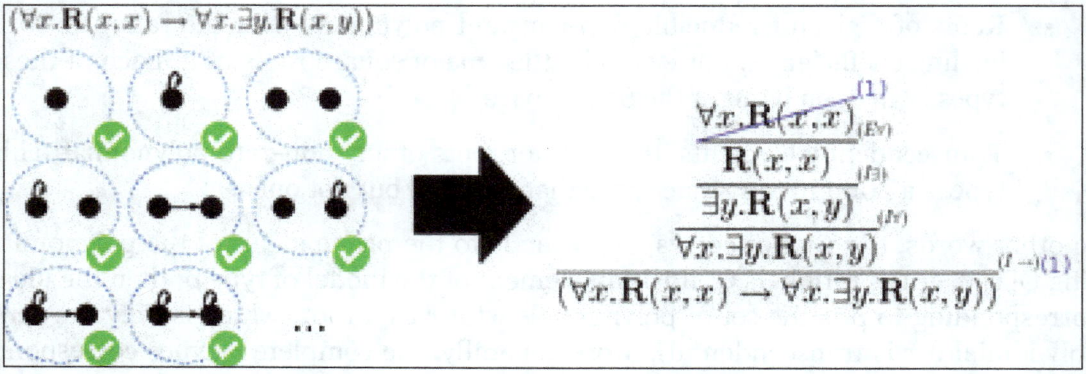

The formula (∀x. R(x, x)) → (∀x∃y. R(x, y)) holds in all structures (only the simplest 8 are shown left). By Gödel›s completeness result, it must hence have a natural deduction proof (shown right).

Preliminaries

There are numerous deductive systems for first-order logic, including systems of natural deduction and Hilbert-style systems. Common to all deductive systems is the notion of a formal deduction. This is a sequence (or, in some cases, a finite tree) of formulas with a specially-designated conclusion. The definition of a deduction is such that it is finite and that it is possible to verify algorithmically (by a computer, for example, or by hand) that a given sequence (or tree) of formulas is indeed a deduction.

A first-order formula is called logically valid if it is true in every structure for the language of the formula (i.e. for any assignment of values to the variables of the formula). To formally state, and then prove, the completeness theorem, it is necessary to also define a deductive system. A deductive system is called complete if every logically valid formula is the conclusion of some formal deduction, and the completeness theorem for a particular deductive system is the theorem that it is complete in this sense. Thus, in a sense, there is a different completeness theorem for each deductive system. A converse to completeness is soundness, the fact that only logically valid formulas are provable in the deductive system.

If some specific deductive system of first-order logic is sound and complete, then it is "perfect" (a formula is provable if and only if it is logically valid), thus equivalent to any other deductive system with the same quality (any proof in one system can be converted into the other).

Statement of the Theorem

We first fix a deductive system of first-order predicate calculus, choosing any of the well-known equivalent systems. Gödel's original proof assumed the Hilbert-Ackermann proof system.

Gödel's Original Formulation

The completeness theorem says that if a formula is logically valid then there is a finite deduction (a formal proof) of the formula.

Thus, the deductive system is "complete" in the sense that no additional inference rules are required to prove all the logically valid formulas. A converse to completeness is soundness, the fact that only logically valid formulas are provable in the deductive system. Together with soundness (whose verification is easy), this theorem implies that a formula is logically valid if and only if it is the conclusion of a formal deduction.

More General Form

The theorem can be expressed more generally in terms of logical consequence. We say that a sentence *s* is a syntactic consequence of a theory *T*, denoted $T \vdash s$, if *s* is provable from *T* in our deductive system. We say that *s* is a semantic consequence of *T*, denoted $T \vDash s$, if *s* holds in every model of *T*. The completeness theorem then says that for any first-order theory *T* with a well-orderable language, and any sentence *s* in the language of *T*, if $T \vDash s$, then $T \vdash s$.

Since the converse (soundness) also holds, it follows that $T \vDash s$ iff $T \vdash s$, and thus that syntactic and semantic consequence are equivalent for first-order logic.

This more general theorem is used implicitly, for example, when a sentence is shown to be provable from the axioms of group theory by considering an arbitrary group and showing that the sentence is satisfied by that group. Gödel's original formulation is deduced by taking the particular case of a theory without any axiom.

Model Existence Theorem

The completeness theorem can also be understood in terms of consistency, as a consequence of Henkin's model existence theorem. We say that a theory *T* is syntactically consistent if there is no sentence *s* such that both *s* and its negation $\neg s$ are provable from *T* in our deductive system. The model existence theorem says that for any first-order theory *T* with a well-orderable language, if *T* is syntactically consistent, then *T* has a model.

Another version, with connections to the Löwenheim–Skolem theorem, says:

Every syntactically consistent, countable first-order theory has a finite or countable model.

Given Henkin's theorem, the completeness theorem can be proved as follows: If $T \vDash s$, then $T \cup \neg s$ does not have models. By the contrapositive of Henkin's theorem, then $T \cup \neg s$ is syntactically inconsistent. So a contradiction ($\bot$) is provable from $T \cup \neg s$ in the deductive system. Hence $(T \cup \neg s) \vdash \bot$, and then by the properties of the deductive system, $T \vdash s$.

As a Theorem of Arithmetic

The Model Existence Theorem and its proof can be formalized in the framework of Peano arithmetic. Precisely, we can systematically define a model of any consistent effective first-order theory *T* in Peano arithmetic by interpreting each symbol of *T* by an arithmetical formula whose free variables are the arguments of the symbol. However, the definition expressed by this formula is not recursive.

Consequences

An important consequence of the completeness theorem is that it is possible to recursively enumerate the semantic consequences of any effective first-order theory, by enumerating all the possible formal deductions from the axioms of the theory, and use this to produce an enumeration of their conclusions.

This comes in contrast with the direct meaning of the notion of semantic consequence, that quantifies over all structures in a particular language, which is clearly not a recursive definition.

Also, it makes the concept of "provability," and thus of "theorem," a clear concept that only depends on the chosen system of axioms of the theory, and not on the choice of a proof system.

Relationship to the Second Incompleteness Theorem

Gödel's second incompleteness theorem, another celebrated result, shows that there are inherent limitations in what can be achieved with formal proofs in mathematics. The name for the incompleteness theorem refers to another meaning of *complete*: A theory *T* is complete (or decidable) if for every formula *f* in the language of *T* either $T \vdash f$ or $T \vdash \neg f$.

Gödel's second incompleteness theorem states that in any consistent effective theory *T* containing Peano arithmetic (PA), a formula C_T like $C_T = \neg(0 = 1)$ expressing the consistency of *T* cannot be proven within *T*.

The completeness theorem implies the existence of a model of *T* in which the formula C_T is false. Such a model (precisely, the set of "natural numbers" it contains) is necessarily a non-standard, as it contains the code number of a proof of a contradiction of *T*. But *T* is consistent when viewed from the outside. Thus this code number of a proof of contradiction of *T* must be a non-standard number.

In fact, the model of *any* theory containing PA obtained by the systematic construction of the arithmetical model existence theorem, is *always* non-standard with a non-equivalent provability predicate and a non-equivalent way to interpret its own construction, so that this construction is non-recursive (as recursive definitions would be unambiguous). Also, there is no recursive non-standard model of PA.

Relationship to the Compactness Theorem

The completeness theorem and the compactness theorem are two cornerstones of first-order logic. While neither of these theorems can be proven in a completely effective manner, each one can be effectively obtained from the other.

The compactness theorem says that if a formula φ is a logical consequence of a (possibly infinite) set of formulas Γ then it is a logical consequence of a finite subset of Γ. This is an immediate consequence of the completeness theorem, because only a finite number of axioms from Γ can be mentioned in a formal deduction of φ, and the soundness of the deductive system then implies φ is a logical consequence of this finite set. This proof of the compactness theorem is originally due to Gödel.

Conversely, for many deductive systems, it is possible to prove the completeness theorem as an effective consequence of the compactness theorem.

The ineffectiveness of the completeness theorem can be measured along the lines of reverse mathematics. When considered over a countable language, the completeness and compactness theorems are equivalent to each other and equivalent to a weak form of choice known as weak König's lemma, with the equivalence provable in RCA_0 (a second-order variant of Peano arithmetic restricted to induction over $\sum_1^0$ formulas). Weak König's lemma is provable in ZF, the system of Zermelo–Fraenkel set theory without axiom of choice, and thus the completeness and compactness theorems for countable languages are provable in ZF.

However the situation is different when the language is of arbitrary large cardinality since then, though the completeness and compactness theorems remain provably equivalent to each other in ZF, they are also provably equivalent to a weak form of the axiom of choice known as the ultrafilter lemma. In particular, no theory extending ZF can prove either the completeness or compactness theorems over arbitrary (possibly uncountable) languages without also proving the ultrafilter lemma on a set of same cardinality.

Completeness in other Logics

The completeness theorem is a central property of first-order logic that does not hold for all logics. Second-order logic, for example, does not have a completeness theorem for its standard semantics (but does have the completeness property for Henkin semantics), and the set of logically-valid formulas in second-order logic is not recursively enumerable. The same is true of all higher-order logics. It is possible to produce sound deductive systems for higher-order logics, but no such system can be complete.

Lindström's theorem states that first-order logic is the strongest (subject to certain constraints) logic satisfying both compactness and completeness. A completeness theorem can be proved for modal logic or intuitionistic logic with respect to Kripke semantics.

Proofs

Gödel's original proof of the theorem proceeded by reducing the problem to a special case for formulas in a certain syntactic form, and then handling this form with an *ad hoc* argument.

In modern logic texts, Gödel's completeness theorem is usually proved with Henkin's proof, rather than with Gödel's original proof. Henkin's proof directly constructs a term model for any consistent first-order theory. James Margetson developed a computerized formal proof using the Isabelle theorem prover. Other proofs are also known.

Gödel's Incompleteness Theorem

Gödel's first incompleteness theorem states that all consistent axiomatic formulations of number theory which include Peano arithmetic include undecidable propositions. This answers in the negative Hilbert's problem asking whether mathematics is "complete" (in the sense that every statement in the language of number theory can be either proved or disproved).

The inclusion of Peano arithmetic is needed, since for example Presburger arithmetic is a consistent axiomatic formulation of number theory, but it is decidable.

However, Gödel's first incompleteness theorem also holds for Robinson arithmetic (though Robinson's result came much later and was proved by Robinson).

Gerhard Gentzen showed that the consistency and completeness of arithmetic can be proved if transfinite induction is used. However, this approach does not allow proof of the consistency of all mathematics.

References

- Marker, David (2002). Model Theory: An Introduction. Graduate Texts in Mathematics 217. Springer. ISBN 0-387-98760-6

- Model-theory: encyclopediaofmath.org, Retrieved 12 May, 2020
- Ahlbrandt, Gisela; Ziegler, Martin (1986), "Quasi finitely axiomatizable totally categorical theories", Annals of Pure and Applied Logic, 30: 63–82, doi:10.1016/0168-0072(86)90037-0
- Goedels-First-Incomplete-ness-Theorem: mathworld.wolfram.com, Retrieved 09 February, 2020
- Poizat, Bruno (2000), A Course in Model Theory, Springer, ISBN 978-0-387-98655-5

4
Proof Theory

Proof theory is the sub-field of mathematical logic which represents proofs as formal mathematical objects for their analysis by mathematical techniques. Some of its concepts are ordinal analysis, reverse mathematics, formal and informal proof, etc. This chapter delves into the concepts related to proof theory for a thorough understanding of it.

A branch of mathematical logic which deals with the concept of a proof in mathematics and with the applications of this concept in various branches of science and technology.

In the wide meaning of the term, a proof is a manner of justification of the validity of some given assertion. To what extent a proof is convincing will mainly depend on the means employed to substantiate the truth. Thus, in exact sciences certain conditions have been established under which a certain experimental fact may be considered to have been proven (constant reproducibility of the experiment, clear description of the experimental technique, the experimental accuracy, the equipment employed, etc.). In mathematics, where the axiomatic method of study is characteristic, the means of proof were sufficiently precisely established at an early stage of its development. In mathematics, a proof consists of a succession of derivations of assertions from previously derived assertions, and the means of such derivation can be exactly analyzed.

The origin of proof theory can be traced to Antiquity (the deductive method of reasoning in elementary geometry, Aristotelian syllogistics, etc.), but the modern stage in its development begins at the turn of the 19th century with the studies of G. Frege, B. Russell, A.N. Whitehead, E. Zermelo, and, in particular, D. Hilbert. At that time, G. Cantor's research in the theory of sets gave rise to antinomies which cast doubt on the validity of even the simplest considerations concerning arbitrary sets. L.E.J. Brouwer severely criticized certain classical methods of proof of the existence of objects in mathematics, and proposed a radical reconstruction of mathematics in the spirit of intuitionism. Problems concerning the foundations of mathematics became especially timely. Hilbert proposed to separate out the part of practical mathematics known as finitary mathematics, which is unobjectionable as regards both the appearance of antinomies and intuitionistic criticism. Finitary mathematics deals solely with constructive objects such as, say, the natural numbers, and with methods of reasoning that agree with the abstraction of potential realizability but are not concerned with the abstraction of actual infinity. In particular, the use of the law of the excluded middle is restricted. In finitary mathematics, no antinomies have been noted and there is no reason for expecting them to appear. Philosophically, the methods of reasoning of finitary mathematics reflect the constructive processes of real activity much more satisfactorily than that in general set-theoretic mathematics. It was the idea of Hilbert to use the solid ground of finitary mathematics as the foundation of all main branches of classical mathematics. He accordingly presented his formalization method, which is one of the basic methods in proof theory.

In general outline, the formalization method may be described as follows. One formulates a logico-mathematic language (an object language) L in terms of which the assertions of a given mathematical theory T are written as formulas. One then describes a certain class A of formulas of L, known as the axioms of the theory, and describes the derivation rules with the aid of which transitions may be made from given formulas to other formulas. The general term postulates applies to both axioms and derivation rules. The formal theory T* (a calculus according to a different terminology) is defined by a description of the postulates. Formulas which are obtainable from the axioms of the formal theory by its derivation rules are said to be deducible or provable in that theory. The deduction process itself may be formulated as a derivation tree. T* is of special interest as regards the contents of the mathematical theory T if the axioms of T* are records of true statements of T and if the derivation rules lead from true statements to true statements. In such a case T* may be considered as a precision of a fragment of T, while the concept of a derivation in T* may be considered as a more precise form of the informal idea of a proof in T, at least within the framework formalized by the calculus T*. Thus, in constructing the calculus T*, one must specify, in the first place, which postulates are to be considered suitable from the point of view of the theory T. However, this does not mean that a developed semantics of T must be available at this stage; rather, it is permissible to employ practical habits, to include the most useful or the most theoretically interesting facts among the postulates, etc. The exact nature of the description of derivations in the calculus T* makes it possible to apply mathematical methods in their study, and thus to give statements on the content and the properties of the theory T.

Proof theory comprises standard methods of formalization of the content of mathematical theories. Axioms and derivation rules of the calculus are usually divided into logical and applied ones. Logical postulates serve to produce statements which are valid by virtue of their form itself, irrespective of the formalized theory. Such postulates define the logic of the formal theory and are formulated in the form of a propositional calculus or predicate calculus. Applied postulates serve to describe truths related to the special features of a given mathematical theory. Examples are: the axiom of choice in axiomatic set theory; the scheme of induction in elementary arithmetic; and bar induction in intuitionistic analysis.

The Hilbert program for the foundations of mathematics may be described as follows.

It may be hoped that any mathematical theory T, no matter how involved or how abstract (e.g. the essential parts of set theory), may be formalized as a calculus T* and that the formulation of the calculus itself requires finitary mathematics alone. Further, by analyzing the conclusions of T* by purely finitary means, one tries to establish the consistency of T* and, consequently, to establish the absence of antinomies in T, at least in that part of it which is reflected in the postulates of T*. It immediately follows, as far as ordinary formalization methods are concerned, that certain very simple statements (in Hilbert's terminology — real statements) are deducible in T* only if they are true in the finitary sense. It was initially hoped that practically all of classical mathematics could be described in a finitary way, after which its consistency could be demonstrated by finitary means. That this program could not be completed was shown in 1931 by K. Gödel, who proved that, on certain natural assumptions, it is not possible to demonstrate the consistency of T* even by the powerful tools formalized in T*. Nevertheless, the study of various formal calculi remains a very important method in the foundations of mathematics. In the first place, it is of interest to construct calculi which reproduce important branches of modern mathematics, with consistency as the guideline, even if it is not yet possible to prove the consistency of such calculi in a manner acceptable to all mathematicians. An example of a calculus of this kind is the Zermelo–Fraenkel system of set theory, in which practically all results of modern set-theoretic mathematics can be deduced. Proofs of non-deducibility, in this theory, of several fundamental hypotheses obtained on the assumption that the theory is consistent indicate that these hypotheses are independent of the set-theoretic principles accepted in mathematics. This in turn may be regarded as a confirmation of the view according to which the existing concepts are insufficient to prove or disprove the hypotheses under consideration. It is in this sense that the independence of Cantor's continuum hypothesis has been established by P. Cohen.

Secondly, extensive study is made of the class of calculi whose consistency can be established by finitary means. Thus Gödel in 1932 proposed a translation converting formulas deducible by classical arithmetical calculus into formulas deducible by intuitionistic arithmetical calculus (i.e. an interpretation of the former calculus into the latter). If the latter is considered consistent (e.g. by virtue of its natural finitary interpretation), it follows that classical arithmetical calculus is self-consistent as well.

Finally, it may be promising to study more extensive methods than Hilbert's traditional finitism which are satisfactory from some other point of view. Thus, remaining in the framework of potential realizability, one may use so-called general inductive definitions. This makes it possible to use semi-formal theories in which some of the derivation rules have an infinite (but constructively generated) set of premises, and to transfer to finitary mathematics many semantic results. This procedure yielded the results obtained by P.S. Novikov, who established the consistency of classical arithmetic using effective functionals of finite type; by C. Spector, who proved the consistency of classical analysis by extending the natural intuitionistic methods of proof to include intuitionistic effective functionals of finite type; and by A.A. Markov, on constructive semantics involving the use of general inductive definitions. In addition, many important problems on calculi may also be considered out of the context of the foundations of mathematics. These include problems of completeness and solvability of formal theories, the problem of independence of certain statements of a given formal theory, etc. It is then unnecessary to impose limitations of definite methods in the reasoning, and it is permissible to develop the theory of proofs as an ordinary mathematical theory, while using any mathematical means of proof that is convincing for the researcher.

A rigorously defined semantics for the formulas of the language under consideration, i.e. a strict definition of the meaning of the statements expressed in that language, serves as an instrument in the study of calculi, and sometimes even as a motivation for the introduction of new calculi. Thus, such a semantics is well-known in classical propositional calculus: Tautologies and only tautologies are deducible in such a calculus. In the general case, in order to prove that a certain formula A is not deducible in a calculus T* under consideration, it is sufficient to construct a semantics for the formulas of the language of this theory such that all deducible formulas in T* are true in this semantics, while A is false. A semantics can be classical, intuitionistic or of some other type, depending on the logical postulates it has to agree with. Non-classical semantics are successfully employed in the study of classical calculi — e.g. Cohen's forcing relationship can naturally be regarded as a modification of intuitionistic semantics. In another variant of Cohen's theory, multi-valued semantics are employed; these are models with truth values in a complete Boolean algebra. On the other hand, semantics of the type of Kripke models, defined by a classical set-theoretic method, make it possible to clarify many properties of modal and non-standard logics, including intuitionistic logic.

Proof theory makes extensive use of algebraic methods in the form of model theory. An algebraic system which brings each initial symbol of the language into correspondence with some algebraic objects forms a natural definition of some classical semantics of the language. An algebraic system is said to be a model of the formal theory T* if all deducible formulas in T* are true in the semantics generated by the algebraic system. Gödel showed in 1931 that any consistent calculus (with a classical logic) has a model. It was independently shown by A.I. Mal'tsev at a later date that if any finite fragment of a calculus has a model, then the calculus as a whole has a model (the so-called compactness theorem of first-order logic). These two theorems form the foundation of an entire trend in mathematical logic.

A survey of non-standard models of arithmetic established that the concept of the natural numbers is not axiomatizable in the framework of a first-order theory, and that the principle of mathematical induction is independent of the other axioms of arithmetical calculus. The relative nature of the concept of the cardinality of a set in classical mathematics was revealed during the study of countable models for formal theories, the interpretation of which is exclusively based on trivially uncountable models (the so-called Skolem's paradox). Many syntactic results were initially obtained from model-theoretic considerations. In terms of constructions of model theory it is possible to give simple criteria for many concepts of interest to proof theory. Thus, according to

Scott's criterion, a class K of algebraic systems of a given language is axiomatizable if and only if it is closed with respect to ultra-products, isomorphisms and taking of elementary subsystems.

A formal theory is said to be decidable if there exists an algorithm which determines for an arbitrary formula A whether or not it is deducible in that theory. It is known that no formal theory containing a certain fragment of the theory of recursive functions is decidable. It follows that elementary arithmetic, the Zermelo–Fraenkel system and many other theories are undecidable as well. Proof theory disposes of powerful methods for the interpretation of theories in terms of other theories; such interpretations may also be used to establish the undecidability of several very simple calculi in which recursive calculi are not interpreted directly. Examples are elementary group theory, the theory of two equivalence relations, an elementary theory of fractional order, etc. On the other hand, examples are available of interesting decidable theories such as elementary geometry, the elementary theory of real numbers, and the theory of sets of natural numbers with a unique successor operation. The decidability of a theory is demonstrated by model-theoretic and syntactic methods. Syntactic methods often yield simpler decidability algorithms. For instance, the decidability of the elementary theory of p-adic numbers was first established by model-theoretic methods. Subsequently a primitive-recursive algorithm was found for the recognition of decidability of this theory by a certain modification of the syntactic method of quantifier elimination. Estimates of the complexity of decidability algorithms of theories are of importance. As a rule, a primitive-recursive solution algorithm is available for decidable theories, and the problem is to find more exact complexity bounds. A promising direction in such studies is the decidability of real fragments of known formal theories. In this connection classical predicate calculus has been studied in much detail, where an effective description has been given of all decidable and undecidable classes of formulas, in terms of the position of quantifiers in the formula and the form of the predicate symbols appearing in the formula. A number of decidable fragments of arithmetical calculus and of elementary set theory have been described.

Methods for estimating the complexity of derivations have attracted the attention of researchers. This area of research comprises problems such as finding relatively short formulas that are derivable in a complex manner, or formulas yielding a large number of results in a relatively simple manner. Such formulas must be regarded as expressing the "depth" of the facts in the theory. Natural measures of the complexity of a proof are studied: the length of the proof; the time needed to find a solution; the complexity of the formulas used in the proof, etc. This is the domain of contact between proof theory and the methods of theoretical cybernetics.

Structural Proof Theory

In mathematical logic, structural proof theory is the subdiscipline of proof theory that studies proof calculi that support a notion of analytic proof, a kind of proof whose semantic properties are exposed. When all the theorems of a logic formalised in a structural proof theory have analytic proofs, then the proof theory can be used to demonstrate such things as consistency, provide decision procedures, and allow mathematical or computational witnesses to be extracted as counterparts to theorems, the kind of task that is more often given to model theory.

Analytic Proof

The notion of analytic proof was introduced into proof theory by Gerhard Gentzen for the sequent calculus; the analytic proofs are those that are cut-free. His natural deduction calculus also supports a notion of analytic proof, as was shown by Dag Prawitz; the definition is slightly more complex—the analytic proofs are the normal forms, which are related to the notion of normal form in term rewriting.

Structures and Connectives

The term structure in structural proof theory comes from a technical notion introduced in the sequent calculus: the sequent calculus represents the judgement made at any stage of an inference using special, extra-logical operators called structural operators: in $A_1,\ldots,A_m \vdash B_1,\ldots,B_n$, the commas to the left of the turnstile are operators normally interpreted as conjunctions, those to the right as disjunctions, whilst the turnstile symbol itself is interpreted as an implication. However, it is important to note that there is a fundamental difference in behaviour between these operators and the logical connectives they are interpreted by in the sequent calculus: the structural operators are used in every rule of the calculus, and are not considered when asking whether the subformula property applies. Furthermore, the logical rules go one way only: logical structure is introduced by logical rules, and cannot be eliminated once created, while structural operators can be introduced and eliminated in the course of a derivation.

The idea of looking at the syntactic features of sequents as special, non-logical operators is not old, and was forced by innovations in proof theory: when the structural operators are as simple as in Getzen's original sequent calculus there is little need to analyse them, but proof calculi of deep inference such as display logic support structural operators as complex as the logical connectives, and demand sophisticated treatment.

Hypersequents

The hypersequent framework extends the ordinary sequent structure to a multiset of sequents, using an additional structural connective | (called the hypersequent bar) to separate different sequents. It has been used to provide analytic calculi for, e.g., modal, intermediate and substructural logics A hypersequent is a structure:

$$\Gamma_1 \vdash \Delta_1 \mid \ldots \mid \Gamma_n \vdash \Delta_n$$

where each $\Gamma_i \vdash \Delta_i$ is an ordinary sequent, called a component of the hypersequent. As for sequents, hypersequents can be based on sets, multisets, or sequences, and the components can be single-conclusion or multi-conclusion sequents. The formula interpretation of the hypersequents depends on the logic under consideration, but is nearly always some form of disjunction. The most common interpretations are as a simple disjunction:

$$(\bigwedge\Gamma_1 \rightarrow \bigvee\Delta_1)\vee\ldots\vee(\bigwedge\Gamma_n \rightarrow \bigvee\Delta_n)$$

For intermediate logics, or as a disjunction of boxes:

$$\Box(\bigwedge\Gamma_1 \rightarrow \bigvee\Delta_1)\vee\ldots\vee\Box(\bigwedge\Gamma_n \rightarrow \bigvee\Delta_n)$$

for modal logics.

In line with the disjunctive interpretation of the hypersequent bar, essentially all hypersequent calculi include the external structural rules, in particular the external weakening rule:

$$\frac{\Gamma_1 \vdash \Delta_1 \,|\ldots|\, \Gamma_n \vdash \Delta_n}{\Gamma_1 \vdash \Delta_1 \,|\ldots|\, \Gamma_n \vdash \Delta_n \,|\, \Sigma \vdash \Pi}$$

And the external contraction rule:

$$\frac{\Gamma_1 \vdash \Delta_1 \,|\ldots|\, \Gamma_n \vdash \Delta_n \,|\, \Gamma_n \vdash \Delta_n}{\Gamma_1 \vdash \Delta_1 \,|\ldots|\, \Gamma_n \vdash \Delta_n}$$

The additional expressivity of the hypersequent framework is provided by rules manipulating the hypersequent structure. An important example is provided by the modalised splitting rule:

$$\frac{\Gamma_1 \vdash \Delta_1 \,|\ldots|\, \Gamma_n \vdash \Delta_n \,|\, \Box\Sigma, \Omega \vdash \Box\Pi, \Theta}{\Gamma_1 \vdash \Delta_1 \,|\ldots|\, \Gamma_n \vdash \Delta_n \,|\, \Box\Sigma \vdash \Box\Pi \,|\, \Omega \vdash \Theta}$$

For modal logic S5, where $\Box\Sigma$ means that every formula in $\Box\Sigma$ is of the form $\Box A$.

Another example is given by the communication rule for intermediate logic LC:

$$\frac{\Gamma_1 \vdash \Delta_1 \,|\ldots|\, \Gamma_n \vdash \Delta_n \,|\, \Omega \vdash A \qquad \Sigma_1 \vdash \Pi_1 \,|\ldots|\, \Sigma_m \vdash \Pi_m \,|\, \Theta \vdash B}{\Gamma_1 \vdash \Delta_1 \,|\ldots|\, \Gamma_n \vdash \Delta_n \,|\, \Sigma_1 \vdash \Pi_1 \,|\ldots|\, \Sigma_m \vdash \Pi_m \,|\, \Omega \vdash B \,|\, \Theta \vdash A}$$

Note that in the communication rule the components are single-conclusion sequents.

Hilbert System

In logic, especially mathematical logic, a Hilbert system, sometimes called Hilbert calculus, Hilbert-style deductive system or Hilbert–Ackermann system, is a type of system of formal deduction attributed to Gottlob Frege and David Hilbert. These deductive systems are most often studied for first-order logic, but are of interest for other logics as well.

Most variants of Hilbert systems take a characteristic tack in the way they balance a trade-off between logical axioms and rules of inference. Hilbert systems can be characterised by the choice of a large number of schemes of logical axioms and a small set of rules of inference. Systems of natural deduction take the opposite tack, including many deduction rules but very few or no axiom schemes. The most commonly studied Hilbert systems have either just one rule of inference – modus ponens, for propositional logics – or two – with generalisation, to handle predicate logics, as well – and several infinite axiom schemes. Hilbert systems for propositional modal logics, sometimes called Hilbert-Lewis systems, are generally axiomatised with two additional rules, the necessitation rule and the uniform substitution rule.

A characteristic feature of the many variants of Hilbert systems is that the context is not changed in any of their rules of inference, while both natural deduction and sequent calculus contain some context-changing rules. Thus, if one is interested only in the derivability of tautologies, no hypothetical judgments, then one can formalize the Hilbert system in such a way that its rules of inference contain only judgments of a rather simple form. The same cannot be done with the other two deductions systems: as context is changed in some of their rules of inferences, they cannot be formalized so that hypothetical judgments could be avoided – not even if we want to use them just for proving derivability of tautologies.

Formal Deductions

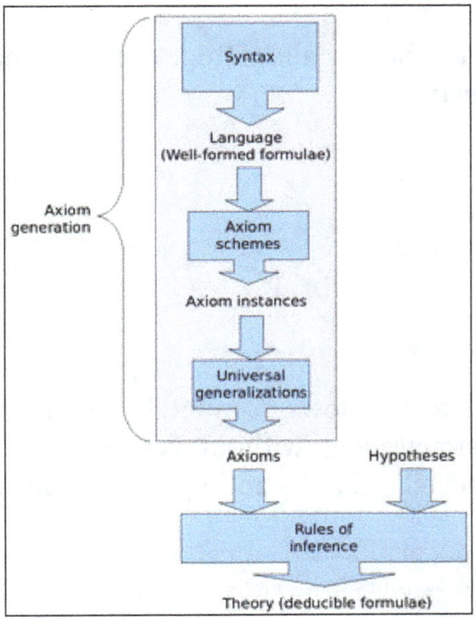

In a Hilbert-style deduction system, a formal deduction is a finite sequence of formulas in which each formula is either an axiom or is obtained from previous formulas by a rule of inference. These formal deductions are meant to mirror natural-language proofs.

Suppose Γ is a set of formulas, considered as hypotheses. For example, Γ could be a set of axioms for group theory or set theory. The notation $\Gamma \vdash \phi$ means that there is a deduction that ends with ϕ using as axioms only logical axioms and elements of Γ. Thus, informally, $\Gamma \vdash \phi$ means that ϕ is provable assuming all the formulas in Γ.

Hilbert-style deduction systems are characterized by the use of numerous schemes of logical axioms. An axiom scheme is an infinite set of axioms obtained by substituting all formulas of some form into a specific pattern. The set of logical axioms includes not only those axioms generated from this pattern, but also any generalization of one of those axioms. A generalization of a formula is obtained by prefixing zero or more universal quantifiers on the formula; for example $\forall y(\forall xPxy \to Pty)$ is a generalization of $\forall xPxy \to Pty$.

Logical Axioms

There are several variant axiomatisations of predicate logic, since for any logic there is freedom in choosing axioms and rules that characterise that logic. We describe here a Hilbert system with nine axioms and just the rule modus ponens, which we call the one-rule axiomatisation and which describes classical equational logic. We deal with a minimal language for this logic, where formulas use only the connectives $\neg$ and $\to$ only the quantifier $\forall$. Later we show how the system can be extended to include additional logical connectives, such as $\wedge$ and $\vee$, without enlarging the class of deducible formulas.

The first four logical axiom schemes allow (together with modus ponens) for the manipulation of logical connectives.

P1: $\phi \to \phi$

P2: $\phi \to (\psi \to \phi)$

P3: (ℵℵℵ̸ℵℵℵ)) (( ) ( ))

P4: $(\neg\phi \to \neg\psi) \to (\psi \to \phi)$

The axiom P1 is redundant, as it follows from P3, P2 and modus ponens. These axioms describe classical propositional logic; without axiom P4 we get positive implicational logic. Minimal logic is achieved either by adding instead the axiom P4m, or by defining $\neg\phi$ and $\phi \to \perp$.

P4m: $(\phi \to \psi) \to ((\phi \to \neg\psi) \to \neg\phi)$

Intuitionistic logic is achieved by adding axioms P4i and P5i to positive implicational logic, or by adding axiom P5i to minimal logic. Both P4i and P5i are theorems of classical propositional logic.

P4i: $(\phi \to \neg\phi) \to \neg\phi$

P5i: $\neg\phi \to (\phi \to \psi)$

Note that these are axiom schemes, which represent infinitely many specific instances of axioms. For example, P1 might represent the particular axiom instance $p \to p$, or it might represent $(p \to q) \to (p \to q)$: the ϕ is a place where any formula can be placed. A variable such as this that ranges over formulae is called a 'schematic variable'.

With a second rule of uniform substitution (US), we can change each of these axiom schemes into a single axiom, replacing each schematic variable by some propositional variable that isn't mentioned in any axiom to get what we call the substitutional axiomatisation. Both formalisations have variables, but where the one-rule axiomatisation has schematic variables that are outside the logic's language, the substitutional axiomatisation uses propositional variables that do the same work by expressing the idea of a variable ranging over formulae with a rule that uses substitution.

Let $\phi(p)$ be a formula with one or more instances of the propositional variable p, and let ϕ be another formula. Then from $\phi(p)$, infer $\phi(\psi)$.

The next three logical axiom schemes provide ways to add, manipulate, and remove universal quantifiers.

Q5: $\forall x(\phi) \to \phi[x := t]$ where t may be substituted for x in ϕ.

Q6: $\forall x(\phi \to \psi) \to (\forall x(\phi) \to \forall x(\psi))$.

Q7: $\phi \to \forall x(\phi)$ where x is not a free variable of ϕ.

These three additional rules extend the propositional system to axiomatise classical predicate logic. Likewise, these three rules extend system for intuitionstic propositional logic (with P1-3 and P4i and P5i) to intuitionistic predicate logic.

Universal quantification is often given an alternative axiomatisation using an extra rule of generalisation, in which case the rules Q6 and Q7 are redundant.

The final axiom schemes are required to work with formulas involving the equality symbol.

I8: $x = x$ for every variable x.

I9: $(x = y) \to (\phi[z := x] \to \phi[z := y])$.

Conservative Extensions

It is common to include in a Hilbert-style deduction system only axioms for implication and negation. Given these axioms, it is possible to form conservative extensions of the deduction theorem that permit the use of additional connectives. These extensions are called conservative because if a formula φ involving new connectives is rewritten as a logically equivalent formula θ involving only negation, implication, and universal quantification, then φ is derivable in the extended system if and only if θ is derivable in the original system. When fully extended, a Hilbert-style system will resemble more closely a system of natural deduction.

Existential Quantification

- Introduction: $\forall x(\phi \to \exists y(\phi[x := y]))$.
- Elimination: $\forall x(\phi \to \psi) \to \exists x(\phi) \to \psi$ where x is not a free variable of ψ.

Conjunction and Disjunction

- Conjunction introduction and elimination:
  - Introduction: $\alpha \to \beta \to \alpha \wedge \beta$.
  - Elimination left: $\alpha \wedge \beta \to \alpha$.
  - Elimination right: $\alpha \wedge \beta \to \beta$.
- Disjunction introduction and elimination:
  - Introduction left: $\alpha \to \alpha \vee \beta$.
  - Introduction right: $\beta \to \alpha \vee \beta$.
  - Elimination: $(\alpha \to \gamma) \to (\beta \to \gamma) \to \alpha \vee \beta \to \gamma$.

Metatheorems

Because Hilbert-style systems have very few deduction rules, it is common to prove metatheorems that show that additional deduction rules add no deductive power, in the sense that a deduction using the new deduction rules can be converted into a deduction using only the original deduction rules.

Some common metatheorems of this form are:

- The deduction theorem: $\Gamma;\phi \vdash \psi$ if and only $\Gamma \vdash \phi \to \psi$.
- $\Gamma \vdash \phi \leftrightarrow \psi$ if and only if $\Gamma \vdash \phi \to \psi$ and $\Gamma \vdash \psi \to \phi$.

- Contraposition: If $\Gamma;\phi \vdash \psi$ then $\Gamma;\neg\psi \vdash \neg\phi$.
- Generalization: If $\Gamma \vdash \phi$ and x does not occur free in any formula of Γ then $\Gamma \vdash \forall x \phi$.

Natural Deduction

In logic and proof theory, natural deduction is a kind of proof calculus in which logical reasoning is expressed by inference rules closely related to the "natural" way of reasoning. This contrasts with Hilbert-style systems, which instead use axioms as much as possible to express the logical laws of deductive reasoning.

Judgments and Propositions

A judgment is something that is knowable, that is, an object of knowledge. It is evident if one in fact knows it. Thus "it is raining" is a judgment, which is evident for the one who knows that it is actually raining; in this case one may readily find evidence for the judgment by looking outside the window or stepping out of the house. In mathematical logic however, evidence is often not as directly observable, but rather deduced from more basic evident judgments. The process of deduction is what constitutes a proof; in other words, a judgment is evident if one has a proof for it.

The most important judgments in logic are of the form "A is true". The letter A stands for any expression representing a proposition; the truth judgments thus require a more primitive judgment: "A is a proposition". Many other judgments have been studied; for example, "A is false", "A is true at time t", "A is necessarily true" or "A is possibly true", "the program M has type τ", "A is achievable from the available resources", and many others. To start with, we shall concern ourselves with the simplest two judgments "A is a proposition" and "A is true", abbreviated as "A prop" and "A true" respectively.

The judgment "A prop" defines the structure of valid proofs of A, which in turn defines the structure of propositions. For this reason, the inference rules for this judgment are sometimes known as formation rules. To illustrate, if we have two propositions A and B (that is, the judgments "A prop" and "B prop" are evident), then we form the compound proposition A and B, written symbolically as "$A \wedge B$". We can write this in the form of an inference rule:

$$\frac{A \text{ prop} \qquad B \text{ prop}}{(A \wedge B) \text{ prop}} \wedge_F$$

Where the parentheses are omitted to make the inference rule more succinct:

$$\frac{A \text{ prop} \qquad B \text{ prop}}{A \wedge B \text{ prop}} \wedge_F$$

This inference rule is schematic: A and B can be instantiated with any expression. The general form of an inference rule is:

$$\frac{J_1 \quad J_2 \quad \cdots \quad J_n}{J}\ \text{name}$$

Where each J_i is a judgment and the inference rule is named "name". The judgments above the line are known as premises, and those below the line are conclusions. Other common logical propositions are disjunction ($A \vee B$), negation ($\neg A$), implication ($A \supset B$), and the logical constants truth ($\top$) and falsehood ($\bot$). Their formation rules are below.

$$\frac{A\ \text{prop} \qquad B\ \text{prop}}{A \vee B\ \text{prop}}\ \vee_F \qquad \frac{A\ \text{prop} \qquad B\ \text{prop}}{A \supset B\ \text{prop}}\ \supset_F$$

$$\frac{}{\top\ \text{prop}}\ \top_F \qquad \frac{}{\bot\ \text{prop}}\ \bot_F \qquad \frac{A\ \text{prop}}{\neg A\ \text{prop}}\ \neg_F$$

Introduction and Elimination

Now we discuss the "A true" judgment. Inference rules that introduce a logical connective in the conclusion are known as introduction rules. To introduce conjunctions, i.e., to conclude "A and B true" for propositions A and B, one requires evidence for "A true" and "B true". As an inference rule:

$$\frac{A\ \text{true} \qquad B\ \text{true}}{(A \wedge B)\ \text{true}}\ \wedge_I$$

It must be understood that in such rules the objects are propositions. That is, the above rule is really an abbreviation for:

$$\frac{A\ \text{prop} \qquad B\ \text{prop} \qquad A\ \text{true} \qquad B\ \text{true}}{(A \wedge B)\ \text{true}}\ \wedge_I$$

This can also be written:

$$\frac{A \wedge B\ \text{prop} \qquad A\ \text{true} \qquad B\ \text{true}}{(A \wedge B)\ \text{true}}\ \wedge_I$$

In this form, the first premise can be satisfied by the $\wedge_F$ formation rule, giving the first two premises of the previous form. In the nullary case, one can derive truth from no premises.

$$\frac{}{\top\ \text{true}}\ \top_I$$

If the truth of a proposition can be established in more than one way, the corresponding connective has multiple introduction rules.

$$\frac{A \text{ true}}{A \vee B \text{ true}} \vee_{I1} \qquad \frac{B \text{ true}}{A \vee B \text{ true}} \vee_{I2}$$

Note that in the nullary case, i.e., for falsehood, there are no introduction rules. Thus one can never infer falsehood from simpler judgments.

Dual to introduction rules are elimination rules to describe how to deconstruct information about a compound proposition into information about its constituents. Thus, from "A ∧ B true", we can conclude "A true" and "B true":

$$\frac{A \wedge B \text{ true}}{A \text{ true}} \wedge_{E1} \qquad \frac{A \wedge B \text{ true}}{B \text{ true}} \wedge_{E2}$$

As an example of the use of inference rules, consider commutativity of conjunction. If A ∧ B is true, then B ∧ A is true; This derivation can be drawn by composing inference rules in such a fashion that premises of a lower inference match the conclusion of the next higher inference.

$$\frac{\dfrac{A \wedge B \text{ true}}{B \text{ true}} \wedge_{E2} \qquad \dfrac{A \wedge B \text{ true}}{A \text{ true}} \wedge_{E1}}{B \wedge A \text{ true}} \wedge_{I}$$

The inference figures we have seen so far are not sufficient to state the rules of implication introduction or disjunction elimination; for these, we need a more general notion of hypothetical derivation.

Hypothetical Derivations

A pervasive operation in mathematical logic is reasoning from assumptions. For example, consider the following derivation:

$$\frac{\dfrac{A \wedge (B \wedge C) \text{ true}}{B \wedge C \text{ true}} \wedge_{E2}}{B \text{ true}} \wedge_{E1}$$

This derivation does not establish the truth of B as such; rather, it establishes the following fact:

If A ∧ (B ∧ C) is true then B is true.

In logic, one says "assuming A ∧ (B ∧ C) is true, we show that B is true"; in other words, the judgment "B true" depends on the assumed judgment "A ∧ (B ∧ C) true". This is a hypothetical derivation, which we write as follows:

$$\begin{array}{c} A \wedge (B \wedge C) \text{ true} \\ \vdots \\ B \text{ true} \end{array}$$

The interpretation is: "B true is derivable from A ∧ (B ∧ C) true". Of course, in this specific example we actually know the derivation of "B true" from "A ∧ (B ∧ C) true", but in general we may not a priori know the derivation. The general form of a hypothetical derivation is:

$$\begin{array}{c} D_1 \quad D_2 \cdots D_n \\ \vdots \\ J \end{array}$$

Each hypothetical derivation has a collection of antecedent derivations (the Di) written on the top line, and a succedent judgment (J) written on the bottom line. Each of the premises may itself be a hypothetical derivation. (For simplicity, we treat a judgment as a premise-less derivation.)

The notion of hypothetical judgment is internalised as the connective of implication. The introduction and elimination rules are as follows:

$$\dfrac{\begin{array}{c} \overline{A \text{ true}}^{\,u} \\ \vdots \\ B \text{ true} \end{array}}{A \supset B \text{ true}} \supset_{I^u} \qquad \dfrac{A \supset B \text{ true} \quad A \text{ true}}{B \text{ true}} \supset_E$$

In the introduction rule, the antecedent named u is discharged in the conclusion. This is a mechanism for delimiting the scope of the hypothesis: its sole reason for existence is to establish "B true"; it cannot be used for any other purpose, and in particular, it cannot be used below the introduction. As an example, consider the derivation of "A ⊃ (B ⊃ (A ∧ B)) true":

$$\dfrac{\dfrac{\dfrac{\overline{A \text{ true}}^{\,u} \quad \overline{B \text{ true}}^{\,w}}{A \wedge B \text{ true}} \wedge_I}{B \supset (A \wedge B) \text{ true}} \supset_{I^w}}{A \supset (B \supset (A \wedge B)) \text{ true}} \supset_{I^u}$$

This full derivation has no unsatisfied premises; however, sub-derivations are hypothetical. For instance, the derivation of "B ⊃ (A ∧ B) true" is hypothetical with antecedent "A true" (named u).

With hypothetical derivations, we can now write the elimination rule for disjunction:

$$\dfrac{A \vee B\ \text{true} \qquad \begin{array}{c}\overline{A\ \text{true}}^{\,u}\\ \vdots \\ C\ \text{true}\end{array} \qquad \begin{array}{c}\overline{B\ \text{true}}^{\,w}\\ \vdots\\ C\ \text{true}\end{array}}{C\ \text{true}} \vee_{E^{u,w}}$$

In words, if A ∨ B is true, and we can derive “C true” both from “A true” and from “B true”, then C is indeed true. Note that this rule does not commit to either “A true” or “B true”. In the zero-ary case, i.e. for falsehood, we obtain the following elimination rule:

$$\dfrac{\bot\ \text{true}}{C\ \text{true}} \bot_E$$

This is read as: if falsehood is true, then any proposition C is true.

Negation is similar to implication.

$$\dfrac{\begin{array}{c}\overline{A\ \text{true}}^{\,u}\\ \vdots \\ p\ \text{true}\end{array}}{\neg A\ \text{true}} \neg_{I^{u,p}} \qquad \dfrac{\neg A\ \text{true} \quad A\ \text{true}}{C\ \text{true}} \neg_E$$

The introduction rule discharges both the name of the hypothesis u, and the succedent p, i.e., the proposition p must not occur in the conclusion A. Since these rules are schematic, the interpretation of the introduction rule is: if from “A true” we can derive for every proposition p that “p true”, then A must be false, i.e., “not A true”. For the elimination, if both A and not A are shown to be true, then there is a contradiction, in which case every proposition C is true. Because the rules for implication and negation are so similar, it should be fairly easy to see that not A and A ⊃ ⊥ are equivalent, i.e., each is derivable from the other.

Consistency, Completeness and Normal Forms

A theory is said to be consistent if falsehood is not provable (from no assumptions) and is complete if every theorem or its negation is provable using the inference rules of the logic. These are statements about the entire logic, and are usually tied to some notion of a model. However, there are local notions of consistency and completeness that are purely syntactic checks on the inference rules, and require no appeals to models. The first of these is local consistency, also known as local reducibility, which says that any derivation containing an introduction of a connective followed immediately by its elimination can be turned into an equivalent derivation without this detour. It is a check on the strength of elimination rules: they must not be so strong that they include knowledge not already contained in their premises. As an example, consider conjunctions.

```
———————— u    ———————— w

A true        B true

—————————————————————————— ∧I ⇒———————— u

                                A true

      A ∧ B true

      ————————————— ∧E1

         A true
```

Dually, local completeness says that the elimination rules are strong enough to decompose a connective into the forms suitable for its introduction rule. Again for conjunctions:

```
————————————— u             —————————————————— u  —————————————————— u

A ∧ B true          ⇒       A ∧ B true            A ∧ B true

                            —————————————— ∧E1    —————————————— ∧E2

                                  A true               B true

                                  ———————————————————————————— ∧I

                                              A ∧ B true
```

These notions correspond exactly to β-reduction (beta reduction) and η-conversion (eta conversion) in the lambda calculus, using the Curry–Howard isomorphism. By local completeness, we see that every derivation can be converted to an equivalent derivation where the principal connective is introduced. In fact, if the entire derivation obeys this ordering of eliminations followed by introductions, then it is said to be normal. In a normal derivation all eliminations happen above introductions. In most logics, every derivation has an equivalent normal derivation, called a normal form. The existence of normal forms is generally hard to prove using natural deduction alone, though such accounts do exist in the literature, most notably by Dag Prawitz in 1961. It is much easier to show this indirectly by means of a cut-free sequent calculus presentation.

First and Higher-order Extensions

Here, we will extend it with a second sort of individuals or terms. More precisely, we will add a new kind of judgment, "t is a term" (or "t term") where t is schematic. We shall fix a countable set V of variables, another countable set F of function symbols, and construct terms with the following formation rules:

$$\frac{v \in V}{v \text{ term}}\ \text{var}_F$$

and,

$$\frac{f \in F \qquad t_1 \text{ term} \qquad t_2 \text{ term} \qquad \cdots \qquad t_n \text{ term}}{f(t_1, t_2, \cdots, t_n) \text{ term}}\ \text{app}_F$$

For propositions, we consider a third countable set P of predicates, and define atomic predicates over terms with the following formation rule:

$$\frac{\phi \in P \qquad t_1 \text{ term} \qquad t_2 \text{ term} \qquad \cdots \qquad t_n \text{ term}}{\phi(t_1, t_2, \cdots, t_n) \text{ prop}}\ \text{pred}_F$$

The first two rules of formation provide a definition of a term that is effectively the same as that defined in term algebra and model theory, although the focus of those fields of study is quite different from natural deduction. The third rule of formation effectively defines an atomic formula, as in first-order logic, and again in model theory.

To these are added a pair of formation rules, defining the notation for quantified propositions; one for universal (∀) and existential (∃) quantification:

$$\frac{x \in V \qquad A \text{ prop}}{\forall x.A \text{ prop}}\ \forall_F \qquad\qquad \frac{x \in V \qquad A \text{ prop}}{\exists x.A \text{ prop}}\ \exists_F$$

The universal quantifier has the introduction and elimination rules:

$$\frac{\begin{array}{c}\overline{a \text{ term}}^{\ u} \\ \vdots \\ [a/x]A \text{ true}\end{array}}{\forall x.A \text{ true}}\ \forall_{I^{u,a}} \qquad\qquad \frac{\forall x.A \text{ true} \qquad t \text{ term}}{[t/x]A \text{ true}}\ \forall_E$$

The existential quantifier has the introduction and elimination rules:

$$\frac{[t/x]A \text{ true}}{\exists x.A \text{ true}}\ \exists_I \qquad\qquad \frac{\exists x.A \text{ true} \qquad \begin{array}{c}\overline{a \text{ term}}^{\ u} \quad \overline{[a/x]A \text{ true}}^{\ v} \\ \vdots \\ C \text{ true}\end{array}}{C \text{ true}}\ \exists_{E^{a,u,v}}$$

In these rules, the notation [t/x] A stands for the substitution of t for every (visible) instance of x in A, avoiding capture. As before the superscripts on the name stand for the components that are discharged: the term a cannot occur in the conclusion of ∀I (such terms are known as eigenvariables or parameters), and the hypotheses named u and v in ∃E are localised to the second premise in a hypothetical derivation. Although the propositional logic was decidable, adding the quantifiers makes the logic undecidable.

So far, the quantified extensions are first-order: they distinguish propositions from the kinds of objects quantified over. Higher-order logic takes a different approach and has only a single sort of propositions. The quantifiers have as the domain of quantification the very same sort of propositions, as reflected in the formation rules:

$$\cfrac{\cfrac{}{p \text{ prop}}u \quad \vdots \quad A \text{ prop}}{\forall p.A \text{ prop}}\forall_{F^u} \qquad \cfrac{\cfrac{}{p \text{ prop}}u \quad \vdots \quad A \text{ prop}}{\exists p.A \text{ prop}}\exists_{F^u}$$

It is possible to be in-between first-order and higher-order logics. For example, second-order logic has two kinds of propositions, one kind quantifying over terms, and the second kind quantifying over propositions of the first kind.

Tree-like Presentations

Gentzen's discharging annotations used to internalise hypothetical judgments can be avoided by representing proofs as a tree of sequents $\Gamma \vdash A$ instead of a tree of A true judgments.

Sequential Presentations

Jaśkowski's representations of natural deduction led to different notations such as Fitch-style calculus (or Fitch's diagrams) or Suppes' method, of which Lemmon gave a variant called system L.

- 1940: In a textbook, Quine indicated antecedent dependencies by line numbers in square brackets, anticipating Suppes' 1957 line-number notation.
- 1950: In a textbook, Quine demonstrated a method of using one or more asterisks to the left of each line of proof to indicate dependencies. This is equivalent to Kleene's vertical bars. (It is not totally clear if Quine's asterisk notation appeared in the original 1950 edition or was added in a later edition.)
- 1957: An introduction to practical logic theorem proving in a textbook by Suppes. This indicated dependencies (i.e. antecedent propositions) by line numbers at the left of each line.
- 1963: Stoll uses sets of line numbers to indicate antecedent dependencies of the lines of sequential logical arguments based on natural deduction inference rules.
- 1965: The entire textbook by Lemmon is an introduction to logic proofs using a method based on that of Suppes.

- 1967: In a textbook, Kleene briefly demonstrated two kinds of practical logic proofs, one system using explicit quotations of antecedent propositions on the left of each line, the other system using vertical bar-lines on the left to indicate dependencies.

Proofs and Type Theory

The presentation of natural deduction so far has concentrated on the nature of propositions without giving a formal definition of a proof. To formalise the notion of proof, we alter the presentation of hypothetical derivations slightly. We label the antecedents with proof variables (from some countable set V of variables), and decorate the succedent with the actual proof. The antecedents or hypotheses are separated from the succedent by means of a turnstile (⊢). This modification sometimes goes under the name of localised hypotheses. The following diagram summarises the change.

$$\begin{array}{ccccc} \overline{}\,u_1 & \overline{}\,u_2 & \cdots & \overline{}\,un & \\ J_1 & J_2 & & J_n & \Rightarrow \quad u_1{:}J_1,\ u_2{:}J_2,\ \ldots,\ u_n{:}J_n \vdash J \\ & \vdots & & & \\ & J & & & \end{array}$$

The collection of hypotheses will be written as Γ when their exact composition is not relevant. To make proofs explicit, we move from the proof-less judgment "A true" to a judgment: "π is a proof of (A true)", which is written symbolically as "π : A true". Following the standard approach, proofs are specified with their own formation rules for the judgment "π proof". The simplest possible proof is the use of a labelled hypothesis; in this case the evidence is the label itself.

$$\frac{u \in V}{u \text{ proof}}\ \text{proof-F} \qquad \frac{}{u{:}A \text{ true} \vdash u : A \text{ true}}\ \text{hyp}$$

For brevity, we shall leave off the judgmental label true, i.e., write "Γ ⊢ π : A". Let us re-examine some of the connectives with explicit proofs. For conjunction, we look at the introduction rule ∧I to discover the form of proofs of conjunction: they must be a pair of proofs of the two conjuncts. Thus:

$$\frac{\pi_1 \text{ proof} \qquad \pi_2 \text{ proof}}{(\pi_1, \pi_2) \text{ proof}}\ \text{pair-F} \qquad \frac{\Gamma \vdash \pi_1 : A \qquad \Gamma \vdash \pi_2 : B}{\Gamma \vdash (\pi_1, \pi_2) : A \wedge B}\ \wedge\text{I}$$

The elimination rules $\wedge E_1$ and $\wedge E_2$ select either the left or the right conjunct; thus the proofs are a pair of projections—first (fst) and second (snd).

$$\frac{\pi \text{ proof}}{\text{fst } \pi \text{ proof}}\ \text{fst-F} \qquad \frac{\Gamma \vdash \pi : A \wedge B}{\Gamma \vdash \text{fst } \pi : A}\ \wedge E_1$$

$$\frac{\pi \text{ proof}}{\text{snd } \pi \text{ proof}}\ \text{snd-F} \qquad \frac{\Gamma \vdash \pi : A \wedge B}{\Gamma \vdash \text{snd } \pi : B}\ \wedge E_2$$

For implication, the introduction form localises or binds the hypothesis, written using a λ; this corresponds to the discharged label. In the rule, "Γ, u:A" stands for the collection of hypotheses Γ, together with the additional hypothesis u.

$$\frac{\pi \text{ proof}}{\lambda u.\ \pi \text{ proof}}\ \lambda\text{-F} \qquad \frac{\Gamma, u{:}A \vdash \pi : B}{\Gamma \vdash \lambda u.\ \pi : A \supset B}\ \supset I$$

$$\frac{\pi_1 \text{ proof} \quad \pi_2 \text{ proof}}{\pi_1\ \pi_2 \text{ proof}}\ \text{app-F} \qquad \frac{\Gamma \vdash \pi_1 : A \supset B \quad \Gamma \vdash \pi_2 : A}{\Gamma \vdash \pi_1\ \pi_2 : B}\ \supset E$$

With proofs available explicitly, one can manipulate and reason about proofs. The key operation on proofs is the substitution of one proof for an assumption used in another proof. This is commonly known as a substitution theorem, and can be proved by induction on the depth (or structure) of the second judgment.

Substitution Theorem

If,

$$\Gamma \vdash \pi_1 : A$$

and,

$$\Gamma, u{:}A \vdash \pi_2 : B,$$

then,

$$\Gamma \vdash [\pi_1/u]\ \pi_2 : B$$

So far the judgment "$\Gamma \vdash \pi : A$" has had a purely logical interpretation. In type theory, the logical view is exchanged for a more computational view of objects. Propositions in the logical interpretation are now viewed as types, and proofs as programs in the lambda calculus. Thus the interpretation of "$\pi : A$" is "the program π has type A". The logical connectives are also given a different reading: conjunction is viewed as product ($\times$), implication as the function arrow ($\rightarrow$), etc. The differences are only cosmetic, however. Type theory has a natural deduction presentation in terms of formation, introduction and elimination rules; in fact, the reader can easily reconstruct what is known as simple type theory.

The difference between logic and type theory is primarily a shift of focus from the types (propositions) to the programs (proofs). Type theory is chiefly interested in the convertibility or reducibility of programs. For every type, there are canonical programs of that type which are irreducible; these are known as canonical forms or values. If every program can be reduced to a canonical form, then the type theory is said to be normalising (or weakly normalising). If the canonical form is unique, then the theory is said to be strongly normalising. Normalisability is a rare feature of most non-trivial type theories, which is a big departure from the logical world. (Recall that almost every logical derivation has an equivalent normal derivation.) To sketch the reason: in type theories that admit recursive definitions, it is possible to write programs that never reduce to a value; such looping programs can generally be given any type. In particular, the looping program has type $\perp$, although there is no logical proof of "$\perp$ true". For this reason, the propositions as types; proofs as programs paradigm only works in one direction, if at all: Interpreting a type theory as a logic generally gives an inconsistent logic.

Dependent Type Theory

Like logic, type theory has many extensions and variants, including first-order and higher-order versions. One branch, known as dependent type theory, is used in a number of computer-assisted proof systems. Dependent type theory allows quantifiers to range over programs themselves. These quantified types are written as Π and Σ instead of $\forall$ and $\exists$, and have the following formation rules:

$$\frac{\Gamma \vdash A \text{ type} \qquad \Gamma,\ x{:}A \vdash B \text{ type}}{\Gamma \vdash \Pi x{:}A.\ B \text{ type}}\ \Pi\text{-F} \qquad \frac{\Gamma \vdash A \text{ type} \qquad \Gamma,\ x{:}A \vdash B \text{ type}}{\Gamma \vdash \Sigma x{:}A.\ B \text{ type}}\ \Sigma\text{-F}$$

These types are generalisations of the arrow and product types, respectively, as witnessed by their introduction and elimination rules.

$$\frac{\Gamma,\ x{:}A \vdash \pi : B}{\Gamma \vdash \lambda x.\ \pi : \Pi x{:}A.\ B}\ \Pi\text{I} \qquad \frac{\Gamma \vdash \pi_1 : \Pi x{:}A.\ B \qquad \Gamma \vdash \pi_2 : A}{\Gamma \vdash \pi_1\ \pi_2 : [\pi_2/x]\ B}\ \Pi\text{E}$$

$$\frac{\Gamma \vdash \pi_1 : A \quad \Gamma, x{:}A \vdash \pi_2 : B}{\Gamma \vdash (\pi_1, \pi_2) : \Sigma x{:}A.\ B}\ \Sigma I \qquad \frac{\Gamma \vdash \pi : \Sigma x{:}A.\ B}{\Gamma \vdash \mathrm{fst}\ \pi : A}\ \Sigma E_1 \qquad \frac{\Gamma \vdash \pi : \Sigma x{:}A.\ B}{\Gamma \vdash \mathrm{snd}\ \pi : [\mathrm{fst}\ \pi/x]\ B}\ \Sigma E_2$$

Dependent type theory in full generality is very powerful: it is able to express almost any conceivable property of programs directly in the types of the program. This generality comes at a steep price — either typechecking is undecidable (extensional type theory), or extensional reasoning is more difficult (intensional type theory). For this reason, some dependent type theories do not allow quantification over arbitrary programs, but rather restrict to programs of a given decidable index domain, for example integers, strings, or linear programs.

Since dependent type theories allow types to depend on programs, a natural question to ask is whether it is possible for programs to depend on types, or any other combination. There are many kinds of answers to such questions. A popular approach in type theory is to allow programs to be quantified over types, also known as parametric polymorphism; of this there are two main kinds: if types and programs are kept separate, then one obtains a somewhat more well-behaved system called predicative polymorphism; if the distinction between program and type is blurred, one obtains the type-theoretic analogue of higher-order logic, also known as impredicative polymorphism. Various combinations of dependency and polymorphism have been considered in the literature, the most famous being the lambda cube of Henk Barendregt.

The intersection of logic and type theory is a vast and active research area. New logics are usually formalised in a general type theoretic setting, known as a logical framework. Popular modern logical frameworks such as the calculus of constructions and LF are based on higher-order dependent type theory, with various trade-offs in terms of decidability and expressive power. These logical frameworks are themselves always specified as natural deduction systems, which is a testament to the versatility of the natural deduction approach.

Classical and Modal Logics

For simplicity, the logics presented so far have been intuitionistic. Classical logic extends intuitionistic logic with an additional axiom or principle of excluded middle:

For any proposition p, the proposition $p \vee \neg p$ is true.

This statement is not obviously either an introduction or elimination; indeed, it involves two distinct connectives. Gentzen's original treatment of excluded middle prescribed one of the following three (equivalent) formulations, which were already present in analogous forms in the systems of Hilbert and Heyting:

$$\frac{}{A \vee \neg A\ \mathtt{true}}\ XM_1 \qquad \frac{\neg\neg A\ \mathtt{true}}{A\ \mathtt{true}}\ XM_2 \qquad \frac{\begin{array}{c}\overline{\neg A\ \mathtt{true}}\ u \\ \vdots \\ p\ \mathtt{true}\end{array}}{A\ \mathtt{true}}\ XM_3 u, p$$

(XM_3 is merely XM_2 expressed in terms of E.) This treatment of excluded middle, in addition to being objectionable from a purist's standpoint, introduces additional complications in the definition of normal forms.

A comparatively more satisfactory treatment of classical natural deduction in terms of introduction and elimination rules alone was first proposed by Parigot in 1992 in the form of a classical lambda calculus called λμ. The key insight of his approach was to replace a truth-centric judgment A true with a more classical notion, reminiscent of the sequent calculus: in localised form, instead of $\Gamma \vdash A$, he used $\Gamma \vdash \Delta$, with Δ a collection of propositions similar to Γ. Γ was treated as a conjunction, and Δ as a disjunction. This structure is essentially lifted directly from classical sequent calculi, but the innovation in λμ was to give a computational meaning to classical natural deduction proofs in terms of a callcc or a throw/catch mechanism seen in LISP and its descendants.

Another important extension was for modal and other logics that need more than just the basic judgment of truth. These were first described, for the alethic modal logics S4 and S5, in a natural deduction style by Prawitz in 1965, and have since accumulated a large body of related work. To give a simple example, the modal logic S4 requires one new judgment, "A valid", that is categorical with respect to truth:

If "A true" under no assumptions of the form "B true", then "A valid".

This categorical judgment is internalised as a unary connective $\Box A$ (read "necessarily A") with the following introduction and elimination rules:

$$\frac{A\ \mathtt{valid}}{\Box A\ \mathtt{true}}\ \Box I \qquad \frac{\Box A\ \mathtt{true}}{A\ \mathtt{true}}\ \Box E$$

Note that the premise "A valid" has no defining rules; instead, the categorical definition of validity is used in its place. This mode becomes clearer in the localised form when the hypotheses are explicit. We write "$\Omega;\Gamma \vdash$ A true" where Γ contains the true hypotheses as before, and Ω contains valid hypotheses. On the right there is just a single judgment "A true"; validity is not needed here since "$\Omega \vdash$ A valid" is by definition the same as "$\Omega;\cdot \vdash$ A true". The introduction and elimination forms are then:

$$\frac{\Omega;\cdot \vdash \pi : \text{A true}}{\Omega;\cdot \vdash \text{box } \pi : \Box \text{ A true}}\ \Box\text{I} \qquad \frac{\Omega;\Gamma \vdash \pi : \Box \text{ A true}}{\Omega;\Gamma \vdash \text{unbox } \pi : \text{A true}}\ \Box\text{E}$$

The modal hypotheses have their own version of the hypothesis rule and substitution theorem.

$$\frac{}{\Omega,\ u\text{: (A valid) ; } \Gamma \vdash u : \text{A true}}\ \text{valid-hyp}$$

Modal Substitution Theorem

If $\Omega;\cdot \vdash \pi_1$: A true and Ω, u: (A valid) ; $\Gamma \vdash \pi_2$: C true, then $\Omega;\Gamma \vdash [\pi_1/u]\ \pi_2$: C true.

This framework of separating judgments into distinct collections of hypotheses, also known as multi-zoned or polyadic contexts is very powerful and extensible; it has been applied for many different modal logics, and also for linear and other substructural logics, to give a few examples. However, relatively few systems of modal logic can be formalised directly in natural deduction. To give proof-theoretic characterisations of these systems, extensions such as labelling or systems of deep inference.

The addition of labels to formulae permits much finer control of the conditions under which rules apply, allowing the more flexible techniques of analytic tableaux to be applied, as has been done in the case of labelled deduction. Labels also allow the naming of worlds in Kripke semantics; Simpson presents an influential technique for converting frame conditions of modal logics in Kripke semantics into inference rules in a natural deduction formalisation of hybrid logic. Stouppa surveys the application of many proof theories, such as Avron and Pottinger's hypersequents and Belnap's display logic to such modal logics as S5 and B.

Comparison with other Foundational Approaches

Sequent Calculus

In the sequent calculus all inference rules have a purely bottom-up reading. Inference rules can apply to elements on both sides of the turnstile. (To differentiate from natural deduction, here, we use a double arrow $\Rightarrow$ instead of the right tack $\vdash$ for sequents.) The introduction rules of natural deduction are viewed as right rules in the sequent calculus, and are structurally very similar. The elimination rules on the other hand turn into left rules in the sequent calculus. To give an example, consider disjunction; the right rules are familiar:

```
Γ ⇒ A                Γ ⇒ B

------------ ∨R1     ----------
                     ∨R2
Γ ⇒ A ∨ B
                     Γ ⇒ A ∨ B
```

On the left:

```
Γ, u:A ⇒ C      Γ, v:B ⇒ C

------------------------------ ∨L

Γ, w: (A ∨ B) ⇒ C
```

Recall the ∨E rule of natural deduction in localised form:

```
Γ ⊢ A ∨ B    Γ, u:A ⊢ C    Γ, v:B ⊢ C

-------------------------------------------- ∨E

Γ ⊢ C
```

The proposition A ∨ B, which is the succedent of a premise in ∨E, turns into a hypothesis of the conclusion in the left rule ∨L. Thus, left rules can be seen as a sort of inverted elimination rule. This observation can be illustrated as follows:

```
          Natural deduction                        Sequent calculus
   ------ hyp

    |
    | elim. rules                       ------------------------- init
    |                                   ↑             ↑
    ↓                                   |             |
   ---------------------- ↑↓ meet       | left rules  | right rules
                                 ⇒
    ↑                                   |             |
    |                                   conclusion
    | intro. rules
    |
   conclusion
```

In the sequent calculus, the left and right rules are performed in lock-step until one reaches the initial sequent, which corresponds to the meeting point of elimination and introduction rules in natural deduction. These initial rules are superficially similar to the hypothesis rule of natural deduction, but in the sequent calculus they describe a transposition or a handshake of a left and a right proposition:

$$\frac{}{\Gamma,\ u{:}A \Rightarrow A}\ \text{init}$$

The correspondence between the sequent calculus and natural deduction is a pair of soundness and completeness theorems, which are both provable by means of an inductive argument.

Soundness of $\Rightarrow$ wrt. $\vdash$: If $\Gamma \Rightarrow A$, then $\Gamma \vdash A$.

Completeness of $\Rightarrow$ wrt. $\vdash$: If $\Gamma \vdash A$, then $\Gamma \Rightarrow A$.

It is clear by these theorems that the sequent calculus does not change the notion of truth, because the same collection of propositions remains true. Thus, one can use the same proof objects as before in sequent calculus derivations. As an example, consider the conjunctions. The right rule is virtually identical to the introduction rule:

Sequent calculus		Natural deduction
$\dfrac{\Gamma \Rightarrow \pi_1 : A \qquad \Gamma \Rightarrow \pi_2 : B}{\Gamma \Rightarrow (\pi_1, \pi_2) : A \wedge B}\ \wedge R$		$\dfrac{\Gamma \vdash \pi_1 : A \qquad \Gamma \vdash \pi_2 : B}{\Gamma \vdash (\pi_1, \pi_2) : A \wedge B}\ \wedge I$

The left rule, however, performs some additional substitutions that are not performed in the corresponding elimination rules.

Sequent calculus		Natural deduction
$\dfrac{\Gamma,\ u{:}A \Rightarrow \pi : C}{\Gamma,\ v{:}\ (A \wedge B) \Rightarrow [\text{fst } v/u]\ \pi : C}\ \wedge L_1$		$\dfrac{\Gamma \vdash \pi : A \wedge B}{\Gamma \vdash \text{fst } \pi : A}\ \wedge E_1$

$\dfrac{\Gamma,\ u{:}B \Rightarrow \pi : C}{\Gamma,\ v{:}\ (A \wedge B) \Rightarrow [\text{snd } v/u]\ \pi : C}\ \wedge L_2$		$\dfrac{\Gamma \vdash \pi : A \wedge B}{\Gamma \vdash \text{snd } \pi : B}\ \wedge E_2$

The kinds of proofs generated in the sequent calculus are therefore rather different from those of natural deduction. The sequent calculus produces proofs in what is known as the β-normal η-long form, which corresponds to a canonical representation of the normal form of the natural deduction proof. If one attempts to describe these proofs using natural deduction itself, one obtains what is called the intercalation calculus, which can be used to formally define the notion of a normal form for natural deduction.

The substitution theorem of natural deduction takes the form of a structural rule or structural theorem known as cut in the sequent calculus.

Cut (Substitution)

If $\Gamma \Rightarrow \pi_1 : A$ and $\Gamma, u{:}A \Rightarrow \pi_2 : C$, then $\Gamma \Rightarrow [\pi_1/u]\,\pi_2 : C$.

In most well behaved logics, cut is unnecessary as an inference rule, though it remains provable as a meta-theorem; the superfluousness of the cut rule is usually presented as a computational process, known as cut elimination. This has an interesting application for natural deduction; usually it is extremely tedious to prove certain properties directly in natural deduction because of an unbounded number of cases. For example, consider showing that a given proposition is not provable in natural deduction. A simple inductive argument fails because of rules like ∨E or E which can introduce arbitrary propositions. However, we know that the sequent calculus is complete with respect to natural deduction, so it is enough to show this unprovability in the sequent calculus. Now, if cut is not available as an inference rule, then all sequent rules either introduce a connective on the right or the left, so the depth of a sequent derivation is fully bounded by the connectives in the final conclusion. Thus, showing unprovability is much easier, because there are only a finite number of cases to consider, and each case is composed entirely of sub-propositions of the conclusion. A simple instance of this is the global consistency theorem: "$\cdot \vdash \bot$ true" is not provable. In the sequent calculus version, this is manifestly true because there is no rule that can have "$\cdot \Rightarrow \bot$" as a conclusion! Proof theorists often prefer to work on cut-free sequent calculus formulations because of such properties.

Sequent Calculus

Sequent calculus is, in essence, a style of formal logical argumentation where every line of a proof is a conditional tautology (called a sequent by Gerhard Gentzen) instead of an unconditional tautology. Each conditional tautology is inferred from other conditional tautologies on earlier lines in a formal argument according to rules and procedures of inference, giving a better approximation to the style of natural deduction used by mathematicians than David Hilbert's earlier style of formal logic where every line was an unconditional tautology. There may be more subtle distinctions to be made; for example, there may be non-logical axioms upon which all propositions are implicitly dependent. Then sequents signify conditional theorems in a first-order language rather than conditional tautologies.

Sequent calculus is one of several extant styles of proof calculus for expressing line-by-line logical arguments.

- Hilbert style: Every line is an unconditional tautology (or theorem).
- Gentzen style: Every line is a conditional tautology (or theorem) with zero or more conditions on the left.
- Natural deduction: Every (conditional) line has exactly one asserted proposition on the right.
- Sequent calculus: Every (conditional) line has zero or more asserted propositions on the right.

In other words, natural deduction and sequent calculus systems are particular distinct kinds of Gentzen-style systems. Hilbert-style systems typically have a very small number of inference rules, relying more on sets of axioms. Gentzen-style systems typically have very few axioms, if any, relying more on sets of rules.

Gentzen-style systems have significant practical and theoretical advantages compared to Hilbert-style systems. For example, both natural deduction and sequent calculus systems facilitate the elimination and introduction of universal and existential quantifiers so that unquantified logical expressions can be manipulated according to the much simpler rules of propositional calculus. In a typical argument, quantifiers are eliminated, then propositional calculus is applied to unquantified expressions (which typically contain free variables), and then the quantifiers are reintroduced. This very much parallels the way in which mathematical proofs are carried out in practice by mathematicians. Predicate calculus proofs are generally much easier to discover with this approach, and are often shorter. Natural deduction systems are more suited to practical theorem-proving. Sequent calculus systems are more suited to theoretical analysis.

In proof theory and mathematical logic, sequent calculus is a family of formal systems sharing a certain style of inference and certain formal properties. The first sequent calculi systems, LK and LJ, were introduced in 1934/1935 by Gerhard Gentzen as a tool for studying natural deduction in first-order logic (in classical and intuitionistic versions, respectively). Gentzen's so-called "Main Theorem" about LK and LJ was the cut-elimination theorem, a result with far-reaching meta-theoretic consequences, including consistency. Gentzen further demonstrated the power and flexibility of this technique a few years later, applying a cut-elimination argument to give a (transfinite) proof of the consistency of Peano arithmetic, in surprising response to Gödel's incompleteness theorems. Since this early work, sequent calculi, also called Gentzen systems, and the general concepts relating to them, have been widely applied in the fields of proof theory, mathematical logic, and automated deduction.

Hilbert-style Deduction Systems

One way to classify different styles of deduction systems is to look at the form of judgments in the system, i.e., which things may appear as the conclusion of a (sub)proof. The simplest judgment form is used in Hilbert-style deduction systems, where a judgment has the form:

$$B,$$

Where, B is any formula of first-order-logic (or whatever logic the deduction system applies to, e.g., propositional calculus or a higher-order logic or a modal logic). The theorems are those formulae that appear as the concluding judgment in a valid proof. A Hilbert-style system needs no distinction between formulae and judgments; we make one here solely for comparison with the cases that follow.

The price paid for the simple syntax of a Hilbert-style system is that complete formal proofs tend to get extremely long. Concrete arguments about proofs in such a system almost always appeal to the deduction theorem. This leads to the idea of including the deduction theorem as a formal rule in the system, which happens in natural deduction.

Natural Deduction Systems

In natural deduction, judgments have the shape:

$$A_1, A_2, \ldots, A_n \vdash B$$

Where, the A_i's and B are again formulae and $n \geq 0$. Permutations of the A_i's are immaterial. In other words, a judgment consists of a list (possibly empty) of formulae on the left-hand side of a turnstile symbol "$\vdash$", with a single formula on the right-hand side. The theorems are those formulae B such that $\vdash B$ (with an empty left-hand side) is the conclusion of a valid proof. (In some presentations of natural deduction, the A_is and the turnstile are not written down explicitly; instead a two-dimensional notation from which they can be inferred is used.)

The standard semantics of a judgment in natural deduction is that it asserts that whenever A_1, A_2, etc., are all true, B will also be true. The judgments:

$$A_1, \ldots, A_n \vdash B$$

and,

$$\vdash (A_1 \wedge \cdots \wedge A_n) \rightarrow B$$

are equivalent in the strong sense that a proof of either one may be extended to a proof of the other.

Sequent Calculus Systems

Finally, sequent calculus generalizes the form of a natural deduction judgment to:

$$A_1,\ldots,A_n \vdash B_1,\ldots,B_k,$$

a syntactic object called a sequent. The formulas on left-hand side of the turnstile are called the antecedent, and the formulas on right-hand side are called the succedent or consequent; together they are called cedents or sequents. Again, A_i and B_i are formulae, and n and k are nonnegative integers, that is, the left-hand-side or the right-hand-side (or neither or both) may be empty. As in natural deduction, theorems are those B where $\vdash B$ is the conclusion of a valid proof.

The standard semantics of a sequent is an assertion that whenever every A_i is true, at least one B_i will also be true. Thus the empty sequent, having both cedents empty, is false. One way to express this is that a comma to the left of the turnstile should be thought of as an "and", and a comma to the right of the turnstile should be thought of as an (inclusive) "or". The sequents:

$$A_1,\ldots,A_n \vdash B_1,\ldots,B_k$$

and,

$$\vdash (A_1 \wedge \cdots \wedge A_n) \rightarrow (B_1 \vee \cdots \vee B_k)$$

are equivalent in the strong sense that a proof of either one may be extended to a proof of the other.

At first sight, this extension of the judgment form may appear to be a strange complication — it is not motivated by an obvious shortcoming of natural deduction, and it is initially confusing that the comma seems to mean entirely different things on the two sides of the turnstile. However, in a classical context the semantics of the sequent can also (by propositional tautology) be expressed either as:

$$\vdash \neg A_1 \vee \neg A_2 \vee \cdots \vee \neg A_n \vee B_1 \vee B_2 \vee \cdots \vee B_k$$

(at least one of the As is false, or one of the Bs is true) or as:

$$\vdash \neg(A_1 \wedge A_2 \wedge \cdots \wedge A_n \wedge \neg B_1 \wedge \neg B_2 \wedge \cdots \wedge \neg B_k)$$

(It cannot be the case that all of the As are true and all of the Bs are false). In these formulations, the only difference between formulae on either side of the turnstile is that one side is negated. Thus, swapping left for right in a sequent corresponds to negating all of the constituent formulae. This means that a symmetry such as De Morgan's laws, which manifests itself as logical negation on the semantic level, translates directly into a left-right symmetry of sequents — and indeed, the inference rules in sequent calculus for dealing with conjunction (∧) are mirror images of those dealing with disjunction (∨).

Many logicians feel that this symmetric presentation offers a deeper insight in the structure of the logic than other styles of proof system, where the classical duality of negation is not as apparent in the rules.

Distinction between Natural Deduction and Sequent Calculus

Gentzen asserted a sharp distinction between his single-output natural deduction systems (NK and NJ) and his multiple-output sequent calculus systems (LK and LJ). He wrote that the intuitionistic natural deduction system NJ was somewhat ugly. He said that the special role of the excluded middle in the classical natural deduction system NK is removed in the classical sequent calculus system LK. He said that the sequent calculus LJ gave more symmetry than natural deduction NJ in the case of intuitionistic logic, as also in the case of classical logic (LK versus NK). Then he said that in addition to these reasons, the sequent calculus with multiple succedent formulas is intended particularly for his principal theorem ("Hauptsatz").

Origin of Word Sequent

The word "sequent" is taken from the word "Sequenz" in Gentzen's 1934 paper. Kleene makes the following comment on the translation into English: "Gentzen says 'Sequenz', which we translate as 'sequent', because we have already used 'sequence' for any succession of objects, where the German is 'Folge'."

Proving Logical Formulas

Reduction Trees

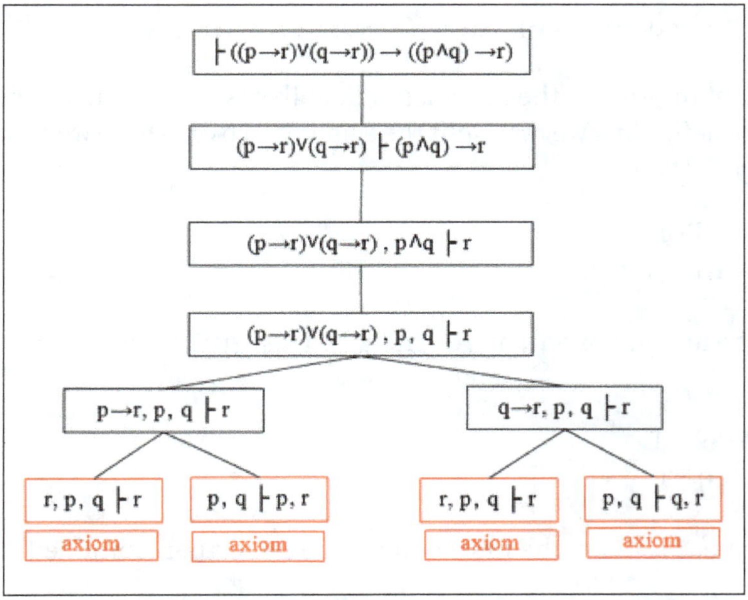

A rooted tree describing a proof finding procedure by sequent calculus.

Sequent calculus can be seen as a tool for proving formulas in propositional logic, similar to the method of analytic tableaux. It gives a series of steps which allows one to reduce the problem of proving a logical formula to simpler and simpler formulas until one arrives at trivial ones.

Consider the following formula:

$$((p \to r) \lor (q \to r)) \to ((p \land q) \to r)$$

This is written in the following form, where the proposition that needs to be proven is to the right of the turnstile symbol $\vdash$:

$$\vdash ((p \to r) \lor (q \to r)) \to ((p \land q) \to r)$$

Now, instead of proving this from the axioms, it is enough to assume the premise of the implication and then try to prove its conclusion. Hence one moves to the following sequent:

$$(p \to r) \lor (q \to r) \vdash (p \land q) \to r$$

Again the right hand side includes an implication, whose premise can further be assumed so that only its conclusion needs to be proven:

$$(p \to r) \lor (q \to r), (p \land q) \vdash r$$

Since the arguments in the left-hand side are assumed to be related by conjunction, this can be replaced by the following:

$$(p \to r) \lor (q \to r), p, q \vdash r$$

This is equivalent to proving the conclusion in both cases of the disjunction on the first argument on the left. Thus we may split the sequent to two, where we now have to prove each separately:

$$p \to r, p, q \vdash r$$
$$q \to r, p, q \vdash r$$

In the case of the first judgment, we rewrite $p \to r$ as $\neg p \lor r$ and split the sequent again to get:

$$\neg p, p, q \vdash r$$
$$r, p, q \vdash r$$

The second sequent is done; the first sequent can be further simplified into:

$$p, q \vdash p, r$$

This process can always be continued until there are only atomic formulas in each side. The root of the tree is the formula we wish to prove; the leaves consist of atomic formulas only. The tree is known as a reduction tree.

The items to the left of the turnstile are understood to be connected by conjunction and those to the right by disjunction. Therefore, when both consist only of atomic symbols, the sequent is provable (and always true) if and only if at least one of the symbols on the right also appears on the left.

Following are the rules by which one proceeds along the tree. Whenever one sequent is split into two, the tree vertex has three edges (one coming from the vertex closer to the root), and the tree is branched. Additionally, one may freely change the order of the arguments in each side; Γ and Δ stand for possible additional arguments.

The usual term for the horizontal line used in Gentzen-style layouts for natural deduction is inference line.

Left	Right
L∧ rule: $\dfrac{\Gamma, A \wedge B \vdash \Delta}{\Gamma, A, B \vdash \Delta}$	R∧ rule: $\dfrac{\Gamma \vdash \Delta, A \wedge B}{\Gamma \vdash \Delta, A \qquad \Gamma \vdash \Delta, B}$
L∨ rule: $\dfrac{\Gamma, A \vee B \vdash \Delta}{\Gamma, A \vdash \Delta \qquad \Gamma, B \vdash \Delta}$	R∨ rule: $\dfrac{\Gamma \vdash \Delta, A \vee B}{\Gamma \vdash \Delta, A, B}$
L→ rule: $\dfrac{\Gamma, A \to B \vdash \Delta}{\Gamma \vdash \Delta, A \qquad \Gamma, B \vdash \Delta}$	R→ rule: $\dfrac{\Gamma \vdash \Delta, A \to B}{\Gamma, A \vdash \Delta, B}$
L¬ rule: $\dfrac{\Gamma, \neg A \vdash \Delta}{\Gamma \vdash \Delta, A}$	R¬ rule: $\dfrac{\Gamma \vdash \Delta, \neg A}{\Gamma, A \vdash \Delta}$

Axiom: $p, r \vdash q, r$

Starting with any formula in propositional logic, by a series of steps, the right side of the turnstile can be processed until it includes only atomic symbols. Then, the same is done for the left side. Since every logical operator appears in one of the rules above, and is omitted by the rule, the process terminates when no logical operators remain: The formula has been decomposed.

Thus, the sequents in the leaves of the trees include only atomic symbols, which are either provable by the axiom or not, according to whether one of the symbols on the right also appears on the left.

It is easy to see that the steps in the tree preserve the semantic truth value of the formulas implied by them, with conjunction understood between the tree's different branches whenever there is a split. It is also obvious that an axiom is provable if and only if it is true for every truth values of the atomic symbols. Thus this system is sound and complete in propositional logic.

Relation to Standard Axiomatizations

Sequent calculus is related to other axiomatizations of propositional calculus, such as Frege's propositional calculus or Jan Łukasiewicz's axiomatization (itself a part of the standard Hilbert system): Every formula that can be proven in these has a reduction tree.

This can be shown as follows: Every proof in propositional calculus uses only axioms and the inference rules. Each use of an axiom scheme yields a true logical formula, and can thus be proven in sequent calculus; examples for these are shown below. The only inference rule in the systems is modus ponens, which is implemented by the cut rule.

The System LK

A (formal) proof in this calculus is a sequence of sequents, where each of the sequents is derivable from sequents appearing earlier in the sequence by using one of the rules below.

Inference Rules

The following notation will be used:

- $\vdash$ known as the turnstile, separates the assumptions on the left from the propositions on the right.
- A and B denote formulae of first-order predicate logic (one may also restrict this to propositional logic).
- Γ, Δ, Σ, and Π are finite (possibly empty) sequences of formulae, called contexts:
    - When on the left of the $\vdash$, the sequence of formulas is considered conjunctively (all assumed to hold at the same time).
    - While on the right of the $\vdash$, the sequence of formulas is considered disjunctively (at least one of the formulas must hold for any assignment of variables).
- t denotes an arbitrary term.
- x and y denote variables.
- a variable is said to occur free within a formula if it occurs outside the scope of quantifiers $\forall$ or $\exists$.

- $A[t/x]$ denotes the formula that is obtained by substituting the term t for every free occurrence of the variable x in formula A with the restriction that the term t must be free for the variable x in A (i.e., no occurrence of any variable in t becomes bound in $A[t/x]$.
- WL and WR stand for Weakening Left/Right, CL and CR for Contraction, and PL and PR for Permutation.
- Note that, contrary to the rules for proceeding along the reduction tree presented above, the following rules are for moving in the opposite directions, from axioms to theorems. Thus they are exact mirror-images of the rules above, except that here symmetry is not implicitly assumed, and rules regarding quantification are added.

Note that, contrary to the rules for proceeding along the reduction tree presented above, the following rules are for moving in the opposite directions, from axioms to theorems. Thus they are exact mirror-images of the rules above, except that here symmetry is not implicitly assumed, and rules regarding quantification are added.

Axiom	Cut
$\dfrac{}{A \vdash A}\ (I)$	$\dfrac{\Gamma \vdash \Delta, A \qquad A, \Sigma \vdash \Pi}{\Gamma, \Sigma \vdash \Delta, \Pi}\ (Cut)$
Left logical rules	Right logical rules
$\dfrac{\Gamma, A \vdash \Delta}{\Gamma, A \wedge B \vdash \Delta}\ (\wedge L_1)$	$\dfrac{\Gamma \vdash A, \Delta}{\Gamma \vdash A \vee B, \Delta}\ (\vee R_1)$
$\dfrac{\Gamma, B \vdash \Delta}{\Gamma, A \wedge B \vdash \Delta}\ (\wedge L_2)$	$\dfrac{\Gamma \vdash B, \Delta}{\Gamma \vdash A \vee B, \Delta}\ (\vee R_2)$
$\dfrac{\Gamma, A \vdash \Delta \qquad \Sigma, B \vdash \Pi}{\Gamma, \Sigma, A \vee B \vdash \Delta, \Pi}\ (\vee L)$	$\dfrac{\Gamma \vdash A, \Delta \qquad \Sigma \vdash B, \Pi}{\Gamma, \Sigma \vdash A \wedge B, \Delta, \Pi}\ (\wedge R)$
$\dfrac{\Gamma \vdash A, \Delta \qquad \Sigma, B \vdash \Pi}{\Gamma, \Sigma, A \to B \vdash \Delta, \Pi}\ (\to L)$	$\dfrac{\Gamma, A \vdash B, \Delta}{\Gamma \vdash A \to B, \Delta}\ (\to R)$
$\dfrac{\Gamma \vdash A, \Delta}{\Gamma, \neg A \vdash \Delta}\ (\neg L)$	$\dfrac{\Gamma, A \vdash \Delta}{\Gamma \vdash \neg A, \Delta}\ (\neg R)$
$\dfrac{\Gamma, A[t/x] \vdash \Delta}{\Gamma, \forall x A \vdash \Delta}\ (\forall L)$	$\dfrac{\Gamma \vdash A[y/x], \Delta}{\Gamma \vdash \forall x A, \Delta}\ (\forall R)$
$\dfrac{\Gamma, A[y/x] \vdash \Delta}{\Gamma, \exists x A \vdash \Delta}\ (\exists L)$	$\dfrac{\Gamma \vdash A[t/x], \Delta}{\Gamma \vdash \exists x A, \Delta}\ (\exists R)$

Left structural rules	Right structural rules
$\dfrac{\Gamma \vdash \Delta}{\Gamma, A \vdash \Delta}$ (WL)	$\dfrac{\Gamma \vdash \Delta}{\Gamma \vdash A, \Delta}$ (WR)
$\dfrac{\Gamma, A, A \vdash \Delta}{\Gamma \quad \vdash \Delta}$ (CL)	$\dfrac{\Gamma \vdash A, A, \Delta}{\Gamma \vdash A, \Delta}$ (CR)
$\dfrac{\Gamma_1, A, B, \Gamma_2 \vdash \Delta}{\Gamma_1, B, A, \Gamma_2 \vdash \Delta}$ (PL)	$\dfrac{\Gamma \vdash \Delta_1, A, B, \Delta_2}{\Gamma \vdash \Delta_1, B, A, \Delta_2}$ (PR)

Restrictions: In the rules $(\forall R)$ and $(\exists L)$, the variable y must not occur free anywhere in the respective lower sequents.

An Intuitive Explanation

The above rules can be divided into two major groups: logical and structural ones. Each of the logical rules introduces a new logical formula either on the left or on the right of the turnstile $\vdash$. In contrast, the structural rules operate on the structure of the sequents, ignoring the exact shape of the formulae. The two exceptions to this general scheme are the axiom of identity (I) and the rule of (Cut).

Although stated in a formal way, the above rules allow for a very intuitive reading in terms of classical logic. Consider, for example, the rule $(\wedge L_1)$. It says that, whenever one can prove that Δ can be concluded from some sequence of formulae that contain A, then one can also conclude Δ from the (stronger) assumption that $A \wedge B$ holds. Likewise, the rule $(\neg R)$ states that, if Γ and A suffice to conclude Δ, then from Γ alone one can either still conclude Δ or A must be false, i.e. $\neg A$ holds. All the rules can be interpreted in this way.

For an intuition about the quantifier rules, consider the rule $(\forall R)$. Of course concluding that $\forall x A$ holds just from the fact that $A[y/x]$ is true is not in general possible. If, however, the variable y is not mentioned elsewhere (i.e. it can still be chosen freely, without influencing the other formulae), then one may assume, that $A[y/x]$ holds for any value of y. The other rules should then be pretty straightforward.

Instead of viewing the rules as descriptions for legal derivations in predicate logic, one may also consider them as instructions for the construction of a proof for a given statement. In this case the rules can be read bottom-up; for example, $(\wedge R)$ says that, to prove that $A \wedge B$ follows from the assumptions Γ and Σ, it suffices to prove that A can be concluded from Γ and B can be concluded from Σ, respectively. Note that, given some antecedent, it is not clear how this is to be split into Γ and Σ. However, there are only finitely many possibilities to be checked since the antecedent by assumption is finite. This also illustrates how proof theory can be viewed as operating on proofs in a combinatorial fashion: given proofs for both A and B, one can construct a proof for $A \wedge B$.

When looking for some proof, most of the rules offer more or less direct recipes of how to do this. The rule of cut is different: it states that, when a formula A can be concluded and this formula may also serve as a premise for concluding other statements, then the formula A can be "cut out" and the respective derivations are joined. When constructing a proof bottom-up, this creates the problem of guessing A (since it does not appear at all below). The cut-elimination theorem is thus crucial to the applications of sequent calculus in automated deduction: it states that all uses of the cut rule can be eliminated from a proof, implying that any provable sequent can be given a cut-free proof.

The second rule that is somewhat special is the axiom of identity (I). The intuitive reading of this is obvious: every formula proves itself. Like the cut rule, the axiom of identity is somewhat redundant: the completeness of atomic initial sequents states that the rule can be restricted to atomic formulas without any loss of provability.

Observe that all rules have mirror companions, except the ones for implication. This reflects the fact that the usual language of first-order logic does not include the "is not implied by" connective $\not\leftarrow$ that would be the De Morgan dual of implication. Adding such a connective with its natural rules would make the calculus completely left-right symmetric.

Example Derivations

Here is the derivation of "$\vdash A \vee \neg A$", known as the Law of excluded middle.

$$\frac{}{A \vdash A}\ (I)$$

$$\frac{A \vdash A}{\vdash \neg A, A}\ (\neg R)$$

$$\frac{\vdash \neg A, A}{\vdash A \vee \neg A, A}\ (\vee R_2)$$

$$\frac{\vdash A \vee \neg A, A}{\vdash A, A \vee \neg A}\ (PR)$$

$$\frac{\vdash A, A \vee \neg A}{\vdash A \vee \neg A, A \vee \neg A}\ (\vee R_1)$$

$$\frac{\vdash A \vee \neg A, A \vee \neg A}{\vdash A \vee \neg A}\ (CR)$$

Next is the proof of a simple fact involving quantifiers. Note that the converse is not true, and its falsity can be seen when attempting to derive it bottom-up, because an existing free variable cannot be used in substitution in the rules $(\forall R)$ and $(\exists L)$.

$$\frac{}{p(x,y)\vdash p(x,y)}\,(\mathrm{I})$$

$$\frac{}{\forall \text{ü(üüüüü)}\vdash}\,(\forall \mathrm{L})$$

$$\frac{}{\forall x\big(p(x,y)\big)\vdash \exists y\big(p(x,y)\big)}\,(\exists \mathrm{R})$$

$$\frac{}{\exists y\Big(\forall x\big(p(x,y)\big)\Big)\vdash \exists y\big(p(x,y)\big)}\,(\exists \mathrm{L})$$

$$\frac{}{\exists y\Big(\forall x\big(p(x,y)\big)\Big)\vdash \forall x\Big(\exists y\big(p(x,y)\big)\Big)}\,(\forall \mathrm{R})$$

For something more interesting we shall prove: $\Big(\big(A\to(B\vee C)\big)\to\Big(\big((B\to\neg A)\wedge\neg C\big)\to\neg A\Big)\Big)$. It is straightforward to find the derivation, which exemplifies the usefulness of LK in automated proving.

$$\frac{}{B\vdash B}\,(\mathrm{I}) \qquad \frac{}{C\vdash C}\,(\mathrm{I})$$

$$\frac{}{B\vee C\vdash B,C}\,(\vee \mathrm{L})$$

$$\frac{}{B\vee C\vdash C,B}\,(\mathrm{PR})$$

$$\frac{}{B\vee C,\neg C\vdash B}\,(\neg \mathrm{L}) \qquad \frac{}{\neg A\vdash \neg A}\,(\mathrm{I})$$

$$\frac{}{(B\vee C),\neg C,(B\to\neg A)\vdash \neg A}\,(\to \mathrm{L})$$

$$\frac{}{(B\vee C),\neg C,\big((B\to\neg A)\wedge\neg C\big)\vdash \neg A}\,(\wedge \mathrm{L}_1)$$

$$\frac{}{(B\vee C),\big((B\to\neg A)\wedge\neg C\big),\neg C\vdash \neg A}\,(\mathrm{PL})$$

$$\frac{}{A\vdash A}\,(\mathrm{I}) \qquad \frac{}{(B\vee C),\big((B\to\neg A)\wedge\neg C\big),\big((B\to\neg A)\wedge\neg C\big)\vdash \neg A}\,(\wedge \mathrm{L}_2)$$

$$\frac{}{\vdash \neg A,A}\,(\neg \mathrm{R}) \qquad \frac{}{(B\vee C),\big((B\to\neg A)\wedge\neg C\big)\vdash \neg A}\,(\mathrm{CL})$$

$$\frac{}{\vdash A, \neg A}(PR) \qquad \frac{}{((B \rightarrow \neg A) \wedge \neg C), (B \vee C) \vdash \neg A}(PL)$$

$$\frac{}{((B \rightarrow \neg A) \wedge \neg C), (A \rightarrow (B \vee C)) \vdash \neg A, \neg A}(\rightarrow L)$$

$$\frac{}{((B \rightarrow \neg A) \wedge \neg C), (A \rightarrow (B \vee C)) \vdash \neg A}(CR)$$

$$\frac{}{(A \rightarrow (B \vee C)), ((B \rightarrow \neg A) \wedge \neg C) \vdash \neg A}(PL)$$

$$\frac{}{(A \rightarrow (B \vee C)) \vdash (((B \rightarrow \neg A) \wedge \neg C) \rightarrow \neg A)}(\rightarrow R)$$

$$\frac{}{\vdash ((A \rightarrow (B \vee C)) \rightarrow (((B \rightarrow \neg A) \wedge \neg C) \rightarrow \neg A))}(\rightarrow R)$$

These derivations also emphasize the strictly formal structure of the sequent calculus. For example, the logical rules as defined above always act on a formula immediately adjacent to the turnstile, such that the permutation rules are necessary. Note, however, that this is in part an artifact of the presentation, in the original style of Gentzen. A common simplification involves the use of multisets of formulas in the interpretation of the sequent, rather than sequences, eliminating the need for an explicit permutation rule. This corresponds to shifting commutativity of assumptions and derivations outside the sequent calculus, whereas LK embeds it within the system itself.

Relation to Analytic Tableaux

For certain formulations (i.e. variants) of the sequent calculus, a proof in such a calculus is isomorphic to an upside-down, closed analytic tableau.

Structural Rules

Weakening (W) allows the addition of arbitrary elements to a sequence. Intuitively, this is allowed in the antecedent because we can always restrict the scope of our proof (if all cars have wheels, then it's safe to say that all black cars have wheels); and in the succedent because we can always allow for alternative conclusions (if all cars have wheels, then it's safe to say that all cars have either wheels or wings).

Contraction (C) and Permutation (P) assure that neither the order (P) nor the multiplicity of occurrences (C) of elements of the sequences matters. Thus, one could instead of sequences also consider sets.

The extra effort of using sequences, however, is justified since part or all of the structural rules may be omitted. Doing so, one obtains the so-called substructural logics.

Properties of the System LK

This system of rules can be shown to be both sound and complete with respect to first-order logic, i.e. a statement A follows semantically from a set of premises ($\Gamma \vDash A$) iff the sequent $\Gamma \vdash A$ can be derived by the above rules.

In the sequent calculus, the rule of cut is admissible. This result is also referred to as Gentzen's Hauptsatz ("Main Theorem").

Variants

The above rules can be modified in various ways:

Minor Structural Alternatives

There is some freedom of choice regarding the technical details of how sequents and structural rules are formalized. As long as every derivation in LK can be effectively transformed to a derivation using the new rules and vice versa, the modified rules may still be called LK.

First of all, the sequents can be viewed to consist of sets or multisets. In this case, the rules for permuting and (when using sets) contracting formulae are obsolete.

The rule of weakening will become admissible, when the axiom (I) is changed, such that any sequent of the form $\Gamma, A \vdash A, \Delta$ can be concluded. This means that A proves A in any context. Any weakening that appears in a derivation can then be performed right at the start. This may be a convenient change when constructing proofs bottom-up.

Independent of these one may also change the way in which contexts are split within the rules: In the cases $(\wedge R), (\vee L)$, and $(\rightarrow L)$ the left context is somehow split into Γ and Σ when going upwards. Since contraction allows for the duplication of these, one may assume that the full context is used in both branches of the derivation. By doing this, one assures that no important premises are lost in the wrong branch. Using weakening, the irrelevant parts of the context can be eliminated later.

Absurdity

One can introduce $\bot$, the absurdity constant representing false, with the axiom:

$$\frac{}{\bot \vdash}$$

Or if, weakening is to be an admissible rule, then with the axiom:

$$\frac{}{\Gamma, \bot \vdash \Delta}$$

With $\bot$, negation can be subsumed as a special case of implication, via the definition $\neg A \Leftrightarrow A \rightarrow \bot$.

Substructural Logics

Alternatively, one may restrict or forbid the use of some of the structural rules. This yields a variety of substructural logic systems. They are generally weaker than LK (i.e., they have fewer theorems), and thus not complete with respect to the standard semantics of first-order logic. However, they have other interesting properties that have led to applications in theoretical computer science and artificial intelligence.

Intuitionistic Sequent Calculus: System LJ

Surprisingly, some small changes in the rules of LK suffice to turn it into a proof system for intuitionistic logic. To this end, one has to restrict to sequents with exactly one formula on the right-hand side, and modify the rules to maintain this invariant. For example, $(\vee L)$ is reformulated as follows (where C is an arbitrary formula):

$$\frac{\Gamma, A \vdash C \qquad \Sigma, B \vdash C}{\Gamma, \Sigma, A \vee B \vdash C} \ (\vee L)$$

The resulting system is called LJ. It is sound and complete with respect to intuitionistic logic and admits a similar cut-elimination proof. This can be used in proving disjunction and existence properties.

In fact, the only two rules in LK that need to be restricted to single-formula consequents are $(\rightarrow R)$ and $(\neg R)$ (and the latter can be seen as a special case of the former, via $\bot$ as described). When multi-formula consequents are interpreted as disjunctions, all of the other inference rules of LK are actually derivable in LJ, while the offending rule is:

$$\frac{\Gamma, A \vdash B \vee C}{\Gamma \vdash (A \rightarrow B) \vee C}$$

This amounts to the propositional formula:

$(A \rightarrow (B \vee C)) \rightarrow ((A \rightarrow B) \vee C)$, a classical tautology that is not constructively valid.

Ordinal Analysis

In proof theory, ordinal analysis assigns ordinals (often large countable ordinals) to mathematical theories as a measure of their strength. If theories have the same proof-theoretic ordinal they are often equiconsistent, and if one theory has a larger proof-theoretic ordinal than another it can often prove the consistency of the second theory.

Ordinal analysis concerns true, effective (recursive) theories that can interpret a sufficient portion of arithmetic to make statements about ordinal notations.

The proof-theoretic ordinal of such a theory T is the smallest ordinal that the theory cannot prove is well founded—the supremum of all ordinals α for which there exists a notation o in Kleene's sense such that T proves that o is an ordinal notation. Equivalently, it is the supremum of all ordinals α such that there exists a recursive relation R on ω (the set of natural numbers) that well-orders it with ordinal α and such that T proves transfinite induction of arithmetical statements for R.

Upper Bound

The existence of a recursive ordinal that the theory fails to prove is well-ordered follows from the Σ_1^1 bounding theorem, as the set of natural numbers that an effective theory proves to be ordinal notations is a Σ_1^0 set. Thus the proof-theoretic ordinal of a theory will always be a (countable) recursive ordinal, that is, less than the Church–Kleene ordinal ω_1^{CK}.

Examples:

Theories with Proof-theoretic Ordinal ω

- Q, Robinson arithmetic (although the definition of the proof-theoretic ordinal for such weak theories has to be tweaked).
- PA^-, the first-order theory of the nonnegative part of a discretely ordered ring.

Theories with Proof-theoretic Ordinal ω^2

- RFA, rudimentary function arithmetic.
- $I\Delta_0$, arithmetic with induction on Δo-predicates without any axiom asserting that exponentiation is total.

Theories with Proof-theoretic Ordinal ω^3

- EFA, elementary function arithmetic.
- $I\Delta_0$+ exp, arithmetic with induction on Δo-predicates augmented by an axiom asserting that exponentiation is total.
- RCA^*_0, a second order form of EFA sometimes used in reverse mathematics.
- WKL^*_0, a second order form of EFA sometimes used in reverse mathematics.

Friedman's grand conjecture suggests that much "ordinary" mathematics can be proved in weak systems having this as their proof-theoretic ordinal.

Theories with Proof-theoretic Ordinal Ω^n (for n = 2, 3, ... ω)

- $I\Delta_0$ or EFA augmented by an axiom ensuring that each element of the n-th level $\mathcal{E}^n$ of the Grzegorczyk hierarchy is total.

Theories with Proof-theoretic Ordinal ω^ω

- RCA_0, recursive comprehension.
- WKL_0, weak König's lemma.
- PRA, primitive recursive arithmetic.
- $I\Sigma_1$, arithmetic with induction on Σ_1-predicates.

Theories with Proof-theoretic Ordinal ε_0

- PA, Peano arithmetic (shown by Gentzen using cut elimination).
- ACA_0, arithmetical comprehension.

Theories with Proof-theoretic Ordinal the Feferman–Schütte Ordinal Γ_0

- ATR_0, arithmetical transfinite recursion.
- Martin-Löf type theory with arbitrarily many finite level universes.

This ordinal is sometimes considered to be the upper limit for "predicative" theories.

Theories with Proof-theoretic Ordinal the Bachmann–Howard Ordinal

- ID_1, the theory of inductive definitions.
- KP, Kripke–Platek set theory with the axiom of infinity.
- CZF, Aczel's constructive Zermelo–Fraenkel set theory.
- EON, a weak variant of the Feferman's explicit mathematics system T_0.

The Kripke-Platek or CZF set theories are weak set theories without axioms for the full powerset given as set of all subsets. Instead, they tend to either have axioms of restricted separation and formation of new sets, or they grant existence of certain function spaces (exponentiation) instead of carving them out from bigger relations.

Theories with Larger Proof-theoretic Ordinals

- Π_1^1-CA_0, Π_1^1 comprehension has a rather large proof-theoretic ordinal, which was described by Takeuti in terms of "ordinal diagrams", and which is bounded by $\psi_0(\Omega_\omega)$ in Buchholz's notation. It is also the ordinal of $ID_{<\omega}$, the theory of finitely iterated inductive definitions. And also the ordinal of MLW, Martin-Löf type theory with indexed W-Types Setzer.

- T_0, Feferman's constructive system of explicit mathematics has a larger proof-theoretic ordinal, which is also the proof-theoretic ordinal of the KPi, Kripke–Platek set theory with iterated admissibles and Σ^1_2-AC+BI.
- KPM, an extension of Kripke–Platek set theory based on a Mahlo cardinal, has a very large proof-theoretic ordinal ϑ, which was described by Rathjen.
- MLM, an extension of Martin-Löf type theory by one Mahlo-universe, has an even larger proof-theoretic ordinal $\psi_{\Omega_1}(\Omega_M + \omega)$.

Most theories capable of describing the power set of the natural numbers have proof-theoretic ordinals that are so large that no explicit combinatorial description has yet been given. This includes second-order arithmetic and set theories with powersets including ZF and ZFC. The strength of intuitionistic ZF (IZF) equals that of ZF.

Provability Logic

Provability logic is a modal logic, in which the box (or "necessity") operator is interpreted as 'it is provable that'. The point is to capture the notion of a proof predicate of a reasonably rich formal theory, such as Peano arithmetic.

Examples:

There are a number of provability logics. The basic system is generally referred to as GL (for Gödel-Löb) or L or K4W. It can be obtained by adding the modal version of Löb's theorem to the logic K (or K4).

Namely, the axioms of GL are all tautologies of classical propositional logic plus all formulas of one of the following forms:

- Distribution axiom: $\Box(p \to q) \to (\Box p \to \Box q)$.
- Löb's axiom: $\Box(\Box p \to p) \to \Box p$.

And the rules of inference are:

- Modus ponens: From $p \to q$ and p conclude q.
- Necessitation: From $\vdash p$ conclude $\vdash \Box p$.

Generalizations

Interpretability logics and Japaridze's polymodal logic present natural extensions of provability logic.

Reverse Mathematics

Reverse mathematics is a program in mathematical logic that seeks to determine which axioms are required to prove theorems of mathematics. Its defining method can briefly be described as "going backwards from the theorems to the axioms", in contrast to the ordinary mathematical practice of deriving theorems from axioms. It can be conceptualized as sculpting out necessary conditions from sufficient ones.

The reverse mathematics program was foreshadowed by results in set theory such as the classical theorem that the axiom of choice and Zorn's lemma are equivalent over ZF set theory. The goal of reverse mathematics, however, is to study possible axioms of ordinary theorems of mathematics rather than possible axioms for set theory.

Reverse mathematics is usually carried out using subsystems of second-order arithmetic, where many of its definitions and methods are inspired by previous work in constructive analysis and proof theory. The use of second-order arithmetic also allows many techniques from recursion theory to be employed; many results in reverse mathematics have corresponding results in computable analysis. Recently, higher-order reverse mathematics has been introduced, in which the focus is on subsystems of higher-order arithmetic, and the associated richer language.

General Principles

In reverse mathematics, one starts with a framework language and a base theory—a core axiom system—that is too weak to prove most of the theorems one might be interested in, but still powerful enough to develop the definitions necessary to state these theorems. For example, to study the theorem "Every bounded sequence of real numbers has a supremum" it is necessary to use a base system which can speak of real numbers and sequences of real numbers.

For each theorem that can be stated in the base system but is not provable in the base system, the goal is to determine the particular axiom system (stronger than the base system) that is necessary to prove that theorem. To show that a system S is required to prove a theorem T, two proofs are required. The first proof shows T is provable from S; this is an ordinary mathematical proof along with a justification that it can be carried out in the system S. The second proof, known as a reversal, shows that T itself implies S; this proof is carried out in the base system. The reversal establishes that no axiom system S′ that extends the base system can be weaker than S while still proving T.

Use of Second-order Arithmetic

Most reverse mathematics research focuses on subsystems of second-order arithmetic. The body of research in reverse mathematics has established that weak subsystems of second-order arithmetic suffice to formalize almost all undergraduate-level mathematics. In second-order arithmetic, all objects can be represented as either natural numbers or sets of natural numbers. For example, in order to prove theorems about real numbers, the real numbers can be represented as Cauchy sequences of rational numbers, each of which can be represented as a set of natural numbers.

The axiom systems most often considered in reverse mathematics are defined using axiom schemes called comprehension schemes. Such a scheme states that any set of natural numbers definable by a formula of a given complexity exists. In this context, the complexity of formulas is measured using the arithmetical hierarchy and analytical hierarchy.

The reason that reverses mathematics is not carried out using set theory as a base system is that the language of set theory is too expressive. Extremely complex sets of natural numbers can be defined by simple formulas in the language of set theory (which can quantify over arbitrary sets). In the context of second-order arithmetic, results such as Post's theorem establish a close link between the complexity of a formula and the (non) computability of the set it defines.

Another effect of using second-order arithmetic is the need to restrict general mathematical theorems to forms that can be expressed within arithmetic. For example, second-order arithmetic can express the principle "Every countable vector space has a basis" but it cannot express the principle "Every vector space has a basis". In practical terms, this means that theorems of algebra and combinatorics are restricted to countable structures, while theorems of analysis and topology are restricted to separable spaces. Many principles that imply the axiom of choice in their general form (such as "Every vector space has a basis") become provable in weak subsystems of second-order arithmetic when they are restricted. For example, "every field has an algebraic closure" is not provable in ZF set theory, but the restricted form "every countable field has an algebraic closure" is provable in RCA_0, the weakest system typically employed in reverse mathematics.

Use of Higher-order Arithmetic

A recent strand of higher-order reverse mathematics research, initiated by Ulrich Kohlenbach, focuses on subsystems of higher-order arithmetic. Due to the richer language of higher-order arithmetic, the use of representations (aka 'codes') common in second-order arithmetic is greatly reduced. For example, a continuous function on the Cantor space is just a function that maps binary sequences to binary sequences, and that also satisfies the usual 'epsilon-delta'-definition of continuity.

Higher-order reverse mathematics includes higher-order versions of (second-order) comprehension schemes. Such a higher-order axiom states the existence of a functional that decides the truth or falsity of formulas of a given complexity. In this context, the complexity of formulas is also measured using the arithmetical hierarchy and analytical hierarchy. The higher-order counterparts of the major subsystems of second-order arithmetic generally prove the same second-order sentences (or a large subset) as the original second-order systems. For instance, the base theory of higher-order reverse mathematics, called RCA_0^ω, proves the same sentences as RCA_0, up to language.

Second-order comprehension axioms easily generalize to the higher-order framework. However, theorems expressing the compactness of basic spaces behave quite differently in second- and higher-order arithmetic: on one hand, when restricted to countable covers/the language of second-order arithmetic, the compactness of the unit interval is provable in WKL_0. On the other hand, given uncountable covers/the language of higher-order arithmetic, the compactness of the unit interval is only provable from (full) second-order arithmetic. Other covering lemmas (e.g. due to Lindelöf, Vitali, Besicovitch, etc.) exhibit the same behavior, and many basic properties of the gauge integral are equivalent to the compactness of the underlying space.

The Big Five Subsystems of Second-order Arithmetic

Second-order arithmetic is a formal theory of the natural numbers and sets of natural numbers. Many mathematical objects, such as countable rings, groups, and fields, as well as points in effective Polish spaces, can be represented as sets of natural numbers, and modulo this representation can be studied in second-order arithmetic.

Reverse mathematics makes use of several subsystems of second-order arithmetic. A typical reverse mathematics theorem shows that a particular mathematical theorem T is equivalent to a particular subsystem S of second-order arithmetic over a weaker subsystem B. This weaker system B is known as the base system for the result; in order for the reverse mathematics result to have meaning, this system must not itself be able to prove the mathematical theorem T.

Simpson describes five particular subsystems of second-order arithmetic, which he calls the Big Five that occur frequently in reverse mathematics. In order of increasing strength, these systems are named by the initialisms RCA_0, WKL_0, ACA_0, ATR_0, and Π^1_1-CA_0.

The following table summarizes the "big five" systems and lists the counterpart systems in higher-order arithmetic. The latter generally prove the same second-order sentences (or a large subset) as the original second-order systems.

Sub-system	Stands for	Ordinal	Corresponds roughly to	Comments	Higher-order counterpart
RCA_0	Recursive comprehension axiom	ω^ω	Constructive mathematics (Bishop)	The base theory	RCA^ω_0; proves the same second-order sentences as RCA_0.
WKL_0	Weak Kőnig's lemma	ω^ω	Finitistic reductionism (Hilbert)	Conservative over PRA (resp. RCA_0) for Π^0_2 (resp. Π^1_1) sentences	Fan functional; computes modulus of uniform continuity on $2^{\mathbb{N}}$ for continuous functions.

ACA_0	Arithmetical comprehension axiom	ε_0	Predicativism (Weyl, Feferman)	Conservative over Peano arithmetic for arithmetical sentences	The 'Turing jump' functional $\exists^2$ expresses the existence of a discontinuous function on $\mathbb{R}$.
ATR_0	Arithmetical transfinite recursion	Γ_0	Predicative reductionism (Friedman, Simpson)	Conservative over Feferman's system IR for Π^1_1 sentences	The 'transfinite recursion' functional outputs the set claimed to exist by ATR_0.
Π^1_1-CA_0	Π^1_1 comprehension axiom	$\Psi_0(\Omega_\omega)$	Impredicativism		The Suslin functional S^2 decides Π^1_1-formulas (restricted to second-order parameters).

The subscript $_0$ in these names means that the induction scheme has been restricted from the full second-order induction scheme. For example, ACAo includes the induction axiom $(0 \in X \wedge \forall n(n \in X \rightarrow n + 1 \in X)) \rightarrow \forall n\ n \in X$. This together with the full comprehension axiom of second-order arithmetic implies the full second-order induction scheme given by the universal closure of $(\varphi(0) \wedge \forall n(\varphi(n) \rightarrow \varphi(n+1))) \rightarrow \forall n\ \varphi(n)$ for any second-order formula φ. However ACA_0 does not have the full comprehension axiom, and the subscript $_0$ is a reminder that it does not have the full second-order induction scheme either. This restriction is important: systems with restricted induction have significantly lower proof-theoretical ordinals than systems with the full second-order induction scheme.

The Base System RCA_0

RCAo is the fragment of second-order arithmetic whose axioms are the axioms of Robinson arithmetic, induction for Σ^0_1 formulas, and comprehension for Δ^0_1 formulas.

The subsystem RCA_0 is the one most commonly used as a base system for reverse mathematics. The initials "RCA" stand for "recursive comprehension axiom", where "recursive" means "computable", as in recursive function. This name is used because RCA_0 corresponds informally to "computable mathematics". In particular, any set of natural numbers that can be proven to exist in RCA_0 is computable, and thus any theorem which implies that noncomputable sets exist is not provable in RCA_0. To this extent, RCA_0 is a constructive system, although it does not meet the requirements of the program of constructivism because it is a theory in classical logic including the law of excluded middle.

Despite its seeming weakness (of not proving any non-computable sets exist), RCA_0 is sufficient to prove a number of classical theorems which, therefore, require only minimal logical strength. These theorems are, in a sense, below the reach of the reverse mathematics enterprise because they are already provable in the base system. The classical theorems provable in RCA_0 include:

- Basic properties of the natural numbers, integers, and rational numbers (for example, that the latter form an ordered field).

- Basic properties of the real numbers (the real numbers are an Archimedean ordered field; any nested sequence of closed intervals whose lengths tend to zero has a single point in its intersection; the real numbers are not countable).
- The Baire category theorem for a complete separable metric space (the separability condition is necessary to even state the theorem in the language of second-order arithmetic).
- The intermediate value theorem on continuous real functions.
- The Banach–Steinhaus theorem for a sequence of continuous linear operators on separable Banach spaces.
- A weak version of Gödel's completeness theorem (for a set of sentences, in a countable language, that is already closed under consequence).
- The existence of an algebraic closure for a countable field (but not its uniqueness).
- The existence and uniqueness of the real closure of a countable ordered field.

The first-order part of RCA_0 (the theorems of the system that do not involve any set variables) is the set of theorems of first-order Peano arithmetic with induction limited to Σ^0_1 formulas. It is provably consistent, as is RCA_0, in full first-order Peano arithmetic.

Weak Kőnig's Lemma WKL_0

The subsystem WKL_0 consists of RCA_0 plus a weak form of Kőnig's lemma, namely the statement that every infinite subtree of the full binary tree (the tree of all finite sequences of 0's and 1's) has an infinite path. This proposition, which is known as weak Kőnig's lemma, is easy to state in the language of second-order arithmetic. WKLo can also be defined as the principle of Σ^0_1 separation (given two Σ^0_1 formulas of a free variable n which are exclusive, there is a class containing all n satisfying the one and no n satisfying the other).

The following remark on terminology is in order. The term "weak Kőnig's lemma" refers to the sentence which says that any infinite subtree of the binary tree has an infinite path. When this axiom is added to RCA_0, the resulting subsystem is called WKL_0. A similar distinction between particular axioms, on the one hand, and subsystems including the basic axioms and induction, on the other hand, is made for the stronger subsystems.

In a sense, weak Kőnig's lemma is a form of the axiom of choice (although, as stated, it can be proven in classical Zermelo–Fraenkel set theory without the axiom of choice). It is not constructively valid in some senses of the word constructive.

To show that WKL_0 is actually stronger than (not provable in) RCAo, it is sufficient to exhibit a theorem of WKL_0 which implies that noncomputable sets exist. This is not difficult; WKL_0 implies the existence of separating sets for effectively inseparable recursively enumerable sets.

It turns out that RCA_0 and WKL_0 have the same first-order part, meaning that they prove the same first-order sentences. WKL_0 can prove a good number of classical mathematical results which do not follow from RCA_0, however. These results are not expressible as first-order statements but can be expressed as second-order statements.

The following results are equivalent to weak Kőnig's lemma and thus to WKL_0 over RCA_0:

- The Heine–Borel theorem for the closed unit real interval, in the following sense: every covering by a sequence of open intervals has a finite subcovering.
- The Heine–Borel theorem for complete totally bounded separable metric spaces (where covering is by a sequence of open balls).
- A continuous real function on the closed unit interval (or on any compact separable metric space, as above) is bounded (or: bounded and reaches its bounds).
- A continuous real function on the closed unit interval can be uniformly approximated by polynomials (with rational coefficients).
- A continuous real function on the closed unit interval is uniformly continuous.
- A continuous real function on the closed unit interval is Riemann integrable.
- The Brouwer fixed point theorem (for continuous functions on a finite product of copies of the closed unit interval).
- The separable Hahn–Banach theorem in the form: a bounded linear form on a subspace of a separable Banach space extends to a bounded linear form on the whole space.
- The Jordan curve theorem.
- Gödel's completeness theorem (for a countable language).
- Determinacy for open (or even clopen) games on $\{0,1\}$ of length ω.
- Every countable commutative ring has a prime ideal.
- Every countable formally real field is orderable.
- Uniqueness of algebraic closure (for a countable field).

Arithmetical Comprehension ACA_0

ACA_0 is RCA_0 plus the comprehension scheme for arithmetical formulas (which is sometimes called the "arithmetical comprehension axiom"). That is, ACA_0 allows us to form the set of natural numbers satisfying an arbitrary arithmetical formula (one with no bound set variables, although possibly containing set parameters). Actually, it suffices to add to RCAo the comprehension scheme for Σ_1 formulas in order to obtain full arithmetical comprehension.

The first-order part of ACA_0 is exactly first-order Peano arithmetic; ACA_0 is a conservative extension of first-order Peano arithmetic. The two systems are provably (in a weak system) equiconsistent. ACA_0 can be thought of as a framework of predicative mathematics, although there are predicatively provable theorems that are not provable in ACA_0. Most of the fundamental results about the natural numbers, and many other mathematical theorems, can be proven in this system.

One way of seeing that ACA_0 is stronger than WKLo is to exhibit a model of WKL_0 that doesn't contain all arithmetical sets. In fact, it is possible to build a model of WKL_0 consisting entirely of low sets using the low basis theorem, since low sets relative to low sets are low.

The following assertions are equivalent to ACA_0 over RCA_0:

- The sequential completeness of the real numbers (every bounded increasing sequence of real numbers has a limit).
- The Bolzano–Weierstrass theorem.
- Ascoli's theorem: every bounded equicontinuous sequence of real functions on the unit interval has a uniformly convergent subsequence.
- Every countable commutative ring has a maximal ideal.
- Every countable vector space over the rationals (or over any countable field) has a basis.
- Every countable field has a transcendence basis.
- Kőnig's lemma (for arbitrary finitely branching trees, as opposed to the weak version).
- Various theorems in combinatorics, such as certain forms of Ramsey's theorem.

Arithmetical Transfinite Recursion ATR_0

The system ATR_0 adds to ACA_0 an axiom which states, informally, that any arithmetical functional (meaning any arithmetical formula with a free number variable n and a free class variable X, seen as the operator taking X to the set of n satisfying the formula) can be iterated transfinitely along any countable well ordering starting with any set. ATR_0 is equivalent over ACA_0 to the principle of Σ_1^0 separation. ATR_0 is impredicative, and has the proof-theoretic ordinal Γ_0, the supremum of that of predicative systems.

ATR_0 proves the consistency of ACA_0, and thus by Gödel's theorem it is strictly stronger.

The following assertions are equivalent to ATR_0 over RCA_0:

- Any two countable well orderings are comparable. That is, they are isomorphic or one is isomorphic to a proper initial segment of the other.
- Ulm's theorem for countable reduced Abelian groups.
- The perfect set theorem, which states that every uncountable closed subset of a complete separable metric space contains a perfect closed set.
- Lusin's separation theorem (essentially Σ_1^1 separation).
- Determinacy for open sets in the Baire space.

Π_1^1 Comprehension Π_1^1-CA_0

Π_1^1-CA_0 is stronger than arithmetical transfinite recursion and is fully impredicative. It consists of RCA_0 plus the comprehension scheme for Π_1^1 formulas.

In a sense, Π_1^1-CA_0 comprehension is to arithmetical transfinite recursion (Σ_1^1 separation) as ACA_0 is to weak Kőnig's lemma (Σ_1^0 separation). It is equivalent to several statements of descriptive set theory whose proofs make use of strongly impredicative arguments; this equivalence shows that these impredicative arguments cannot be removed.

The following theorems are equivalent to Π_1^1-CA_0 over RCA_0:

- The Cantor–Bendixson theorem (every closed set of reals is the union of a perfect set and a countable set).
- Every countable abelian group is the direct sum of a divisible group and a reduced group.

Additional Systems

- Weaker systems than recursive comprehension can be defined. The weak system RCA_0^* consists of elementary function arithmetic EFA (the basic axioms plus Δ_0^0 induction in the enriched language with an exponential operation) plus Δ_1^0 comprehension. Over RCA_0^*, recursive comprehension as defined earlier (that is, with Σ_1^0 induction) is equivalent to the statement that a polynomial (over a countable field) has only finitely many roots and to the classification theorem for finitely generated Abelian groups. The system RCA_0^* has the same proof theoretic ordinal ω^3 as EFA and is conservative over EFA for Π_2^0 sentences.
- Weak Weak Kőnig's Lemma is the statement that a subtree of the infinite binary tree having no infinite paths has an asymptotically vanishing proportion of the

leaves at length n (with a uniform estimate as to how many leaves of length n exist). An equivalent formulation is that any subset of Cantor space that has positive measure is nonempty (this is not provable in RCA_0). $WWKL_0$ is obtained by adjoining this axiom to RCA_0. It is equivalent to the statement that if the unit real interval is covered by a sequence of intervals then the sum of their lengths is at least one. The model theory of $WWKL_0$ is closely connected to the theory of algorithmically random sequences. In particular, an ω-model of RCA_0 satisfies weak weak Kőnig's lemma if and only if for every set X there is a set Y which is 1-random relative to X.

- DNR (short for "diagonally non-recursive") adds to RCA_0 an axiom asserting the existence of a diagonally non-recursive function relative to every set. That is, DNR states that, for any set A, there exists a total function f such that for all e the eth partial recursive function with oracle A is not equal to f. DNR is strictly weaker than WWKL.

- Δ^1_1-comprehension is in certain ways analogous to arithmetical transfinite recursion as recursive comprehension is to weak Kőnig's lemma. It has the hyperarithmetical sets as minimal ω-model. Arithmetical transfinite recursion proves Δ^1_1-comprehension but not the other way around.

- Σ^1_1-choice is the statement that if $\eta(n, X)$ is a Σ^1_1 formula such that for each n there exists an X satisfying η then there is a sequence of sets X_n such that $\eta(n, X_n)$ holds for each n. Σ^1_1-choice also has the hyperarithmetical sets as minimal ω-model. Arithmetical transfinite recursion proves Σ^1_1-choice but not the other way around.

- HBU (short for "uncountable Heine-Borel") expresses the (open-cover) compactness of the unit interval, involving uncountable covers. The latter aspect of HBU makes it only expressible in the language of third-order arithmetic. Cousin's theorem implies HBU, and these theorems use the same notion of cover due to Cousin and Lindelöf. HBU is hard to prove: in terms of the usual hierarchy of comprehension axioms, a proof of HBU requires full second-order arithmetic.

ω-models and β-models

The ω in ω-model stands for the set of non-negative integers (or finite ordinals). An ω-model is a model for a fragment of second-order arithmetic whose first-order part is the standard model of Peano arithmetic, but whose second-order part may be non-standard. More precisely, an ω-model is given by a choice $S \subseteq 2^\omega$ of subsets of ω. The first-order variables are interpreted in the usual way as elements of ω, and +, × have their usual meanings, while second-order variables are interpreted as elements of S. There is a standard ω model where one just takes S to consist of all subsets of the integers. However, there are also other ω-models; for example, RCA_0 has a minimal ω-model where S consists of the recursive subsets of ω.

A β model is an ω model that is equivalent to the standard ω-model for Π^1_1 and Σ^1_1 sentences (with parameters).

Non-ω models are also useful, especially in the proofs of conservation theorems.

Formal Proof

In logic and mathematics, a formal proof or derivation is a finite sequence of sentences (called well-formed formulas in the case of a formal language), each of which is an axiom, an assumption, or follows from the preceding sentences in the sequence by a rule of inference. It differs from a natural language argument in that it is rigorous, unambiguous and mechanically checkable. If the set of assumptions is empty, then the last sentence in a formal proof is called a theorem of the formal system. The notion of theorem is not in general effective, therefore there may be no method by which we can always find a proof of a given sentence or determine that none exists. The concepts of Fitch-style proof, sequent calculus and natural deduction are generalizations of the concept of proof.

The theorem is a syntactic consequence of all the well-formed formulas preceding it in the proof. For a well-formed formula to qualify as part of a proof, it must be the result of applying a rule of the deductive apparatus (of some formal system) to the previous well-formed formulas in the proof sequence.

Formal proofs often are constructed with the help of computers in interactive theorem proving (e.g., through the use of proof checker and automated theorem prover). Significantly, these proofs can be checked automatically, also by computer. Checking formal proofs is usually simple, while the problem of finding proofs (automated theorem proving) is usually computationally intractable and only semi-decidable, depending upon the formal system in use.

Informal Proof

Proof can be a strange word, at times. It carries various assumptions and meanings with it, and sometimes it is difficult to discern the exact meaning. There is a need for proof in the study of mathematics. Otherwise we arrive at incorrect conclusions. Mathematical proofs come in a variety of formats, the most fundamental of which is an informal proof. With an informal proof, we might see compelling evidence that something is so but, at this level, it is possible that an exception exists somewhere.

Consider the medians of a triangle as shown in figure. A median is a segment that connects a vertex with the midpoint of the side opposite the vertex. It looks as if the three medians in each of the triangles are concurrent (meet at a common point). If you use a dynamic geometry program and move the vertices of a triangle around, the medians will appear to be concurrent. As compelling as this might be, it is not a proof. We can only prove by example if we do every possible example and, even with a dynamic geometry program, you can't examine all possible triangles. Zoom in really close to a suspected point of concurrency - could there be a tiny triangle formed where the point of concurrency appears to be? In this particular example, we know that will not happen, because a formal proof has been done to show that the medians of any triangle are concurrent.

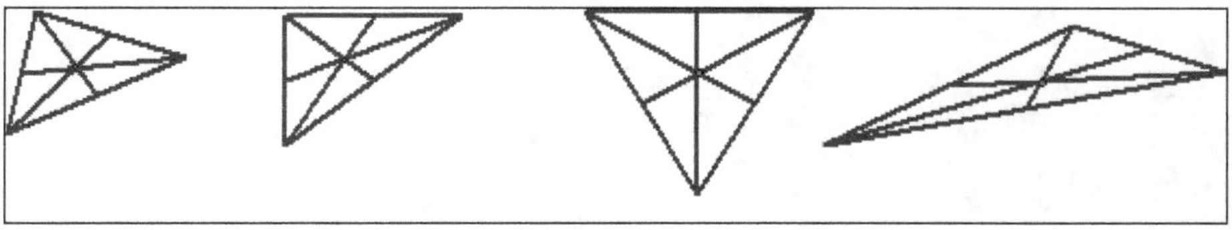

While you are unable to prove that any particular statement is true by using any number of examples (unless you do every possible one), you are able to disprove any statement with one single counterexample. This is not only true for an informal proof, but for any formal proof as well. Proof and disproof by counterexample are powerful devices in a mathematician's tool belt. You can show that the set of digits is not closed for multiplication using a counterexample such as 3 x 4 = 12. A mathematician might write this proof: Suppose the operation of multiplication is closed on the set of digits. Arbitrarily choose 3 and 4 to multiply together, yielding 3 x 4 = 12. Since 12 is not a digit, we have a contradiction. Thus, the operation of multiplication is not closed on the set of digits.

Example: Informally prove or disprove the following statement: The set of whole numbers is closed for multiplication.

Observe the following examples of multiplication using the set of whole numbers:

5 x 7 = 35; 23 x 38 = 874; 125 x 587 = 73,375

Informally prove or disprove the following statement: The set of digits is closed for addition.

Observe the following examples of addition using the set of digits:

1 + 3 = 4; 3 + 5 = 8; 6 + 8 = 14

6 + 8 = 14 and 14 is not an element of the set of digits.

References

- Barker-Plummer, Dave; Barwise, Jon; Etchemendy, John (2011). Language Proof and Logic (2nd ed.). CSLI Publications. ISBN 978-1575866321
- Proof-theory: encyclopediaofmath.org, Retrieved 26 June, 2020

- Ambos-Spies, K.; Kjos-Hanssen, B.; Lempp, S.; Slaman, T.A. (2004), "Comparing DNR and WWKL", Journal of Symbolic Logic, 69 (4): 1089, arXiv:1408.2281, doi:10.2178/jsl/1102022212
- Harrison, John (December 2008). "Formal Proof—Theory and Practice" (PDF). ams.org. Retrieved 2019-12-12
- Kohlenbach, Ulrich (2005), "Higher order reverse mathematics", Higher Order Reverse Mathematics, Reverse Mathematics 2001 (PDF), Lecture notes in Logic, Cambridge University Press, pp. 281–295, doi:10.1017/9781316755846.018, ISBN 9781316755846

5
Formal Logical Systems

Formal logical system refers to the set of inference rules that are used to conclude an expression, axioms and derived theorems. A few of its examples are axiomatic system, formal ethics, Lambda calculus, proof calculus, etc. This chapter sheds light on different formal logical systems to provide an in-depth understanding of the subject.

Formal System

Formal system, also called logistic system, in logic and mathematics are abstract, theoretical organization of terms and implicit relationships that is used as a tool for the analysis of the concept of deduction. Models—structures that interpret the symbols of a formal system—are often used in conjunction with formal systems.

Each formal system has a formal language composed of primitive symbols acted on by certain rules of formation (statements concerning the symbols, functions, and sentences allowable in the system) and developed by inference from a set of axioms. The system thus consists of any number of formulas built up through finite combinations of the primitive symbols—combinations that are formed from the axioms in accordance with the stated rules.

In an axiomatic system, the primitive symbols are undefined; and all other symbols are defined in terms of them. In the Peano postulates for the integers, for example, 0 and $'$ are taken as primitive, and 1 and 2 are defined by $1 = 0'$ and $2 = 1'$. Similarly, in geometry such concepts as "point," "line," and "lies on" are usually posited as primitive terms.

From the primitive symbols, certain formulas are defined as well formed, some of which are listed as axioms; and rules are stated for inferring one formula as a conclusion from one or more other formulas taken as premises. A theorem within such a system is a formula capable of proof through a finite sequence of well-formed formulas, each of which either is an axiom or is inferred from earlier formulas.

A formal system that is treated apart from intended interpretation is a mathematical construct and is more properly called logical calculus; this kind of formulation deals rather with validity and satisfiability than with truth or falsity, which are at the root of formal systems.

In general, then, a formal system provides an ideal language by means of which to abstract and analyze the deductive structure of thought apart from specific meanings. Together with the concept of a model, such systems have formed the basis for a rapidly expanding inquiry into the foundations of mathematics and of other deductive sciences and have even been used to a limited extent in analyzing the empirical sciences.

Formal Logic

Formal logic is the abstract study of propositions, statements, or assertively used sentences and of deductive arguments. The discipline abstracts from the content of these elements the structures or logical forms that they embody. The logician customarily uses a symbolic notation to express such structures clearly and unambiguously and to enable manipulations and tests of validity to be more easily applied.

Formal logic is an a priori, and not an empirical, study. In this respect it contrasts with the natural sciences and with all other disciplines that depend on observation for their data. Its nearest analogy is to pure mathematics; indeed, many logicians and pure mathematicians would regard their respective subjects as indistinguishable, or as merely two stages of the same unified discipline. Formal logic, therefore, is not to be confused with the empirical study of the processes of reasoning, which belongs to psychology. It must also be distinguished from the art of correct reasoning, which is the practical skill of applying logical principles to particular cases; and, even more sharply, it must be distinguished from the art of persuasion, in which invalid arguments are sometimes more effective than valid ones.

Probably the most natural approach to formal logic is through the idea of the validity of an argument of the kind known as deductive. A deductive argument can be roughly characterized as one in which the claim is made that some proposition (the conclusion) follows with strict necessity from some other proposition or propositions (the premises)—i.e., that it would be inconsistent or self-contradictory to assert the premises but deny the conclusion.

If a deductive argument is to succeed in establishing the truth of its conclusion, two quite distinct conditions must be met: first, the conclusion must really follow from the premises—i.e., the deduction of the conclusion from the premises must be logically correct—and, second, the premises themselves must be true. An argument meeting both these conditions is called sound. Of these two conditions, the logician as such is concerned only with the first; the second, the determination of the truth or falsity of the premises, is the task of some special discipline or of common observation appropriate to the subject matter of the argument. When the conclusion of an argument is correctly deducible from its premises, the inference from the premises to the conclusion is said to be (deductively) valid, irrespective of whether the premises are true or false. Other ways of expressing the fact that an inference is deductively valid are to say that the truth of the premises gives (or would give) an absolute guarantee of the truth of the conclusion or that it would involve a logical inconsistency (as distinct from a mere mistake of fact) to suppose that the premises were true but the conclusion false.

The deductive inferences with which formal logic is concerned are, as the name suggests, those for which validity depends not on any features of their subject matter but on their form or structure. Thus, the two inferences:

- Every dog is a mammal. Some quadrupeds are dogs. Therefore, some quadrupeds are mammals.
- Every anarchist is a believer in free love. Some members of the government party are anarchists. Therefore, some members of the government party are believers in free love.

Differ in subject matter and hence require different procedures to check the truth or falsity of their premises. But their validity is ensured by what they have in common—namely, that the argument in each is of the form:

- Every X is a Y. Some Z's are X's. Therefore, some Z's are Y's.

Line above may be called an inference form, and then instances of that inference form. The letters—X, Y, and Z— mark the places into which expressions of a certain type may be inserted. Symbols used for this purpose are known as variables; their use is analogous to that of the x in algebra, which marks the place into which a numeral can be inserted. An instance of an inference form is produced by replacing all the variables in it by appropriate expressions (i.e., ones that make sense in the context) and by doing so uniformly (i.e., by substituting the same expression wherever the same variable recurs). The feature guarantees that every instance of it will be valid is its construction in such a manner that every uniform way of replacing its variables to make the premises true automatically makes the conclusion true also, or, in other words, that no instance of it can have true premises but a false conclusion. In virtue of this feature, the form is termed a valid inference form. In contrast:

- Every X is a Y. Some Z's are Y's. Therefore, some Z's are X's.

is not a valid inference form, for, although instances of it can be produced in which premises and conclusion are all true, instances of it can also be produced in which the premises are true but the conclusion is false—e.g.,

- Every dog is a mammal. Some winged creatures are mammals. Therefore, some winged creatures are dogs.

Formal logic as a study is concerned with inference forms rather than with particular instances of them. One of its tasks is to discriminate between valid and invalid inference forms and to explore and systematize the relations that hold among valid ones.

Closely related to the idea of a valid inference form is that of a valid proposition form. A proposition form is an expression of which the instances (produced as before by appropriate and uniform replacements for variables) are not inferences from several propositions to a conclusion but rather propositions taken individually, and a valid proposition form is one for which all of the instances are true propositions. A simple example is:

- Nothing is both an X and a non-X.

Formal logic is concerned with proposition forms as well as with inference forms. The study of proposition forms can, in fact, be made to include that of inference forms in the following way: let the premises of any given inference form (taken together) be abbreviated by alpha (α) and its conclusion by beta (β). Then the condition stated above for the validity of the inference form "α, therefore β" amounts to saying that no instance of the proposition form "α and not-β" is true—i.e., that every instance of the proposition form:

- Not both: α and not-β

Is true—or that line (7), fully spelled out, of course, is a valid proposition form. The study of proposition forms, however, cannot be similarly accommodated under the study of inference forms, and so for reasons of comprehensiveness it is usual to regard formal logic as the study of proposition forms. Because a logician's handling of proposition forms is in many ways analogous to a mathematician's handling of numerical formulas, the systems he constructs are often called calculi.

Much of the work of a logician proceeds at a more abstract level than that of the foregoing discussion. Even a formula such as above, though not referring to any specific subject matter, contains expressions like "every" and "is a," which are thought of as having a definite meaning, and the variables are intended to mark the places for expressions of one particular kind (roughly, common nouns or class names). It is possible, however—and for some purposes it is essential—to study formulas without attaching even this degree of meaningfulness to them. The construction of a system of logic, in fact, involves two distinguishable processes: one consists in setting up a symbolic apparatus—a set of symbols, rules for stringing these together into formulas, and rules for manipulating these formulas; the second consists in attaching certain meanings to these symbols and formulas. If only the former is done, the system is said to be uninterpreted, or purely formal; if the latter is done as well, the system is said to be interpreted. This distinction is important, because systems of logic turn out to have certain properties quite independently of any interpretations that may be placed upon them. An axiomatic system of logic can be taken as an example—i.e., a system in which certain unproved formulas, known as axioms, are taken as starting points, and further formulas (theorems) are proved on the strength of these. As will appear later, the question whether a sequence of formulas in an axiomatic system is a proof or not depends solely on which formulas are taken as axioms and on what the rules are for deriving theorems from axioms, and not at all on what the theorems or axioms mean. Moreover, a given uninterpreted system is in general capable of being interpreted equally well in a number of different ways; hence, in studying an uninterpreted system, one is studying the structure that is common to a variety of interpreted systems. Normally a logician who constructs a purely formal system does have a particular interpretation in mind, and his motive for constructing it is the belief that when this interpretation is given to it, the formulas of the system will be able to express true principles in some field of thought; but, for the above reasons among others, he will usually take care to describe the formulas and state the rules of the system without reference to interpretation and to indicate as a separate matter the interpretation that he has in mind.

Many of the ideas used in the exposition of formal logic, including some that are mentioned above, raise problems that belong to philosophy rather than to logic itself. Examples are: What is the correct analysis of the notion of truth? What is a proposition, and how is it related to the sentence by which it is expressed? Are there some kinds of sound reasoning that are neither deductive nor inductive? Fortunately, it is possible to learn to do formal logic without having satisfactory answers to such questions, just as it is possible to do mathematics without answering questions belonging to the philosophy of mathematics such as: Are numbers real objects or mental constructs?

Predicate Calculus

Propositions may also be built up, not out of other propositions but out of elements that are not themselves propositions. The simplest kind to be considered here are propositions in which a certain object or individual (in a wide sense) is said to possess a certain property or characteristic; e.g., "Socrates is wise" and "The number 7 is prime." Such a proposition contains two distinguishable parts: (1) an expression that names or designates an individual and (2) an expression, called a predicate that stands for the property that that individual is said to possess. If x, y, z, ... are used as individual variables (replaceable by names of individuals) and the symbols ϕ (phi), ψ (psi), χ (chi), ... as predicate variables (replaceable by predicates), the formula ϕx is used to express the form of the propositions in question. Here x is said to be the argument of ϕ; a predicate (or predicate variable) with only a single argument is said to be a monadic, or one-place, predicate (variable). Predicates with two or more arguments stand not for properties of single individuals but for relations between individuals. Thus the proposition "Tom is a son of John" is analyzable into two names of individuals ("Tom" and "John") and a dyadic or two-place predicate ("is a son of"), of which they are the arguments; and the proposition is thus of the form ϕxy. Analogously, "... is between ... and ..." is a three-place predicate, requiring three arguments, and so on. In general, a predicate variable followed by any number of individual variables is a wff of the predicate calculus. Such a wff is known as an atomic formula, and the predicate variable in it is said to be of degree n, if n is the number of individual variables following it. The degree of a predicate variable is sometimes indicated by a superscript—e.g., ϕxyz may be written as $\phi^3 xyz$; $\phi^3 xy$ would then be regarded as not well formed. This practice is theoretically more accurate, but the superscripts are commonly omitted for ease of reading when no confusion is likely to arise.

Atomic formulas may be combined with truth-functional operators to give formulas such as $\phi x \vee \psi y$ [example: "Either the customer (x) is friendly (ϕ) or else John (y) is disappointed (ψ)"]; $\psi xy \supset \sim\psi x$ [example: "If the road (x) is above (ϕ) the flood line (y), then the road is not wet ($\sim\psi$)"]; and so on. Formulas so formed, however, are valid when and only when they are substitution-instances of valid wffs of PC and hence in a sense do not transcend PC. More interesting formulas are formed by the use, in addition, of quantifiers. There are two kinds of quantifiers: universal quantifiers, written as "(∀ __)" or often simply as "(___)," where the blank is filled by a variable, which may be read, "For all __ "; and existential quantifiers, written as "(∃ __)," which may be read, "For some __ " or "There is a __ such that." ("Some" is to be understood as meaning "at least one.") Thus, $(\forall x)\phi x$ is to mean "For all x, x is ϕ" or, more simply, "Everything is ϕ"; and $(\exists x)\phi x$ is to mean "For some x, x is ϕ" or, more simply, "Something is ϕ" or "There is a ϕ." Slightly more complex examples are $(\forall x)(\phi x \supset \psi x)$ for "Whatever is ϕ is ψ," $(\exists x)(\phi x \cdot \psi x)$ for "Something is both ϕ and ψ," $(\forall x)(\exists y)\phi xy$ for "Everything bears the relation ϕ to at least one thing," and $(\exists x)(\forall y)\phi xy$ for "There is something that bears the relation ϕ to everything." To take a concrete case, if ϕxy means "x loves y" and the values of x and y are taken to be human beings, then the last two formulas mean, respectively, "Everybody loves somebody" and "Somebody loves everybody."

Intuitively, the notions expressed by the words some and every are connected in the following way: to assert that something has a certain property amounts to denying that everything lacks that property (for example, to say that something is white is to say that not everything is nonwhite); and, similarly, to assert that everything has a certain property amounts to denying that there is something that lacks it. These intuitive connections are reflected in the usual practice of taking one of the quantifiers as primitive and defining the other in terms of it. Thus ∀ may be taken as primitive, and ∃ introduced by the definition:

$$(\exists a)\alpha =_{Df} \sim(\forall a)\sim\alpha$$

In which a is any variable and α is any wff; alternatively, ∃ may be taken as primitive, and ∀ introduced by the definition:

$$(\forall a)\alpha =_{Df} \sim(\exists a)\sim\alpha.$$

Lower Predicate Calculus

A predicate calculus in which the only variables that occur in quantifiers are individual variables is known as a lower (or first-order) predicate calculus. Various lower predicate calculi have been constructed. In the most straightforward of these, to which the most attention will be devoted in this discussion and which subsequently will be referred to simply as LPC, the wffs can be specified as follows: Let the primitive symbols be (1) x, y, ... (individual variables), (2) ϕ, ψ, ..., each of some specified degree (predicate variables), and (3) the symbols ~, ∨, ∀, (, and). An infinite number of each type of variable can now be secured as before by the use of numerical subscripts. The symbols ·, ⊃, and ≡ are defined as in PC, and ∃ as explained above. The formation rules are:

- An expression consisting of a predicate variable of degree n followed by n individual variables is a wff.
- If α is a wff, so is ~α.
- If α and β are wffs, so is (α ∨ β).
- If α is a wff and a is an individual variable, then (∀a)α is a wff. (In such a wff, α is said to be the scope of the quantifier.)

If a is any individual variable and α is any wff, every occurrence of a in α is said to be bound (by the quantifiers) when occurring in the wffs (∀a)α and (∃a)α. Any occurrence of a variable that is not bound is said to be free. Thus, in (∀x)(ϕx ∨ ϕy) the x in ϕx is bound, since it occurs within the scope of a quantifier containing x, but y is free. In the wffs of a lower predicate calculus, every occurrence of a predicate variable (ϕ, ψ, χ, ...) is free. A wff containing no free individual variables is said to be a closed wff of LPC. If a wff of LPC is considered as a proposition form, instances of it are obtained by replacing all free variables in it by predicates or by names of individuals, as appropriate. A bound variable, on the other hand, indicates not a point in the wff where a replacement is needed but a point (so to speak) at which the relevant quantifier applies.

For example, in ϕx, in which both variables are free, each variable must be replaced appropriately if a proposition of the form in question (such as "Socrates is wise") is to be obtained; but in (∃x)ϕx, in which x is bound, it is necessary only to replace ϕ by a predicate in order to obtain a complete proposition (e.g., replacing ϕ by "is wise" yields the proposition "Something is wise").

Validity in LPC

Intuitively, a wff of LPC is valid if and only if all its instances are true—i.e., if and only if every result of replacing each of its free variables appropriately and uniformly is a true proposition. A formal definition of validity in LPC to express this intuitive notion more precisely can be given as follows: for any wff of LPC, any number of LPC models can be formed. An LPC model has two elements. One is a set, D, of objects, known as a domain. D may contain as many or as few objects as one chooses, but it must contain at least one, and the objects may be of any kind. The other element, V, is a system of value assignments satisfying the following conditions. To each individual variable there is assigned some member of D (not necessarily a different one in each case). Assignments are next made to the predicate variables in the following way: if ϕ is monadic, there is assigned to it some subset of D (possibly the whole of D); intuitively this subset can be viewed as the set of all the objects in D that have the property ϕ. If ϕ is dyadic, there is assigned to it some set of ordered pairs (i.e., pairs of objects of which one is marked out as the first and the other as the second) drawn from D; intuitively these can be viewed as all the pairs of objects in D in which the relation ϕ holds between the first object in the pair and the second. In general, if ϕ is of degree n, there is assigned to it some set of ordered n-tuples (groups of n objects) of members of D. It is then stipulated that an atomic formula is to have the value 1 in the model if the members of D assigned to its individual variables form, in that order, one of the n-tuples assigned to the predicate variable in it; otherwise, it is to have the value 0. Thus, in the simplest case, ϕx will have the value 1 if the object assigned to x is one object in the set of objects assigned to ϕ; and, if it is not, then ϕx will have the value 0. The values of truth functions are determined by the values of their arguments, as in PC. Finally, the value of $(\forall x)\alpha$ is to be 1 if both (1) the value of α itself is 1 and (2) α would always still have the value 1 if a different assignment were made to x but all the other assignments were left precisely as they were; otherwise $(\forall x)\alpha$ is to have the value 0. Since $\exists$ can be defined in terms of $\forall$, these rules cover all the wffs of LPC. A given wff may of course have the value 1 in some LPC models but the value 0 in others. But a valid wff of LPC may now be defined as one that has the value 1 in every LPC model. If 1 and 0 are viewed as representing truth and falsity, respectively, then validity is defined as truth in every model.

Although the above definition of validity in LPC is quite precise, it does not yield, as did the corresponding definition of PC validity in terms of truth tables, an effective decision procedure. It can, indeed, be shown that no generally applicable decision procedure for LPC is possible—i.e., that LPC is not a decidable system. This does not mean that it is never possible to prove that a given wff of LPC is valid—the validity of an unlimited number of such wffs can in fact be demonstrated—but it does mean that in the case of LPC, unlike that of PC, there is no general procedure, stated in advance, that would enable one to determine, for any wff whatever, whether it is valid or not.

Logical Manipulations in LPC

The intuitive connections between some and every are reflected in the fact that the following equivalences are valid:

$$(\exists x)\phi x \equiv \sim(\forall x)\sim\phi x$$

$$(\forall x)\phi x \equiv \sim(\exists x)\sim\phi x$$

These equivalences remain valid when ϕx is replaced by any wff, however complex; i.e., for any wff α whatsoever,

$$(\exists x)\alpha \equiv \sim(\forall x)\sim\alpha$$

and,

$$(\forall x)(\forall y)\alpha \equiv \sim(\exists x)(\exists y)\sim\alpha$$

and by the resulting replaceability anywhere in a wff of $(\exists x)(\exists y)$ by $\sim(\forall x)(\forall y)\sim$, or of $(\forall x)(\forall y)$ by $\sim(\exists x)(\exists y)\sim$.

Analogously, $(\exists x)(\forall y)$ can be replaced by $\sim(\forall x)(\exists y)\sim$ [e.g., $(\exists x)(\forall y)$(x loves y)—"There is someone who loves everyone"—is equivalent to $\sim(\forall x)(\exists y)\sim$(x loves y)—"It is not true of everyone that there is someone whom he does not love"]; $(\forall x)(\exists y)$ can be replaced by $\sim(\exists x)(\forall y)\sim$; and in general the following rule, covering sequences of quantifiers of any length, holds:

- If a wff contains an unbroken sequence of quantifiers, then the wff that results from replacing $\forall$ by $\exists$ and vice versa throughout that sequence and inserting or deleting $\sim$ at each end of it is equivalent to the original wff.

This may be called the rule of quantifier transformation. It reflects, in a generalized form, the intuitive connections between some and every.

The following are also valid, again where α is any wff:

$$(\forall x)(\forall y)\alpha \equiv (\forall y)(\forall x)\alpha$$

$$(\exists x)(\exists y)\alpha \equiv (\exists y)(\exists x)\alpha$$

The extensions of these lead to the following rule:

- If a wff contains an unbroken sequence either of universal or of existential quantifiers, these quantifiers may be rearranged in any order and the resulting wff will be equivalent to the original wff.

This may be called the rule of quantifier rearrangement.

Two other important rules concern implications, not equivalences:

- If a wff β begins with an unbroken sequence of quantifiers, and β' is obtained from β by replacing $\forall$ by $\exists$ at one or more places in the sequence, then β is stronger than β'—in the sense that $(\beta \supset \beta')$ is valid but $(\beta' \supset \beta)$ is in general not valid.
- If a wff β begins with an unbroken sequence of quantifiers in which some

existential quantifier Q1 precedes some universal quantifier Q2, and if β' is obtained from β by moving Q1 to the right of Q2, then β is stronger than β'.

As illustrations of these rules, the following are valid for any wff α:

$$(\forall x)(\forall y)\alpha \supset (\exists x)(\forall y)\alpha$$

$$(\exists x)(\forall y)(\forall z)\alpha \supset (\exists x)(\forall y)(\exists z)\alpha$$

$$(\exists x)(\forall y)\alpha \supset (\forall y)(\exists x)\alpha$$

$$(\exists x)(\exists y)(\forall z)\alpha \supset (\exists y)(\forall z)(\exists x)\alpha$$

Some of the uses of the above rules can be illustrated by considering a wff α that contains precisely two free individual variables. By prefixing to α two appropriate quantifiers and possibly one or more negation signs, it is possible to form a closed wff (called a closure of α) that will express a determinate proposition when a meaning is assigned to the predicate variables. The above rules can be used to list exhaustively the nonequivalent closures of α and the implication relations between them. The simplest example is ϕxy, which for illustrative purposes can be taken to mean "x loves y." Application of rules 1 and 2 will show that every closure of ϕxy is equivalent to one or another of the following 12 wffs (none of which is in fact equivalent to any of the others):

- $(\forall x)(\forall y)\phi xy$ ("Everybody loves everybody").
- $(\exists x)(\forall y)\phi xy$ ("Somebody loves everybody").
- $(\exists y)(\forall x)\phi xy$ ("There is someone whom everyone loves").
- $(\forall y)(\exists x)\phi xy$ ("Each person is loved by at least one person").
- $(\forall x)(\exists y)\phi xy$ ("Each person loves at least one person").
- $(\exists x)(\exists y)\phi xy$ ("Somebody loves somebody").
- (g)–(l) the respective negations of each of the above.

Rules 3 and 4 show that the following implications among formulas (a)–(f) are valid:

a) ⊃ (b)	(d) ⊃ (f)	(c) ⊃ (e)
(b) ⊃ (d)	(a) ⊃ (c)	(e) ⊃ (f)

The implications holding among the negations of (a)–(f) follow from these by the law of transposition; e.g., since (a) ⊃ (b) is valid, so is ~(b) ⊃ ~(a). The quantification of wffs containing three, four, etc., variables can be dealt with by the same rules.

Intuitively, $(\forall x)\phi x$ and $(\forall y)\phi y$ both "say the same thing"—namely, that everything is ϕ—and $(\exists x)\phi x$ and $(\exists y)\phi y$ both mean simply that something is ϕ. Clearly, so long as the same variable occurs both in the quantifier and as the argument of ϕ, it does not matter what letter is chosen for this purpose. The procedure of replacing some variable in a quantifier, together with every occurrence of that variable in its scope, by some other variable that does not occur elsewhere in its scope is known as relettering a bound variable. If β is the result of relettering a bound variable in a wff α, then α and β are said to be bound alphabetical variants of each other, and bound alphabetical variants are always equivalent. The reason for restricting the replacement variable to one not occurring elsewhere in the scope of the quantifier can be seen from an example: If ϕxy is taken as before to mean "x loves y," the wff $(\forall x)\phi xy$ expresses the proposition form "Everyone loves y," in which the identity of y is left unspecified, and so does its bound alphabetical variant $(\forall z)\phi zy$. If x were replaced by y, however, the closed wff $(\forall y)\phi yy$ would be obtained, which expresses the proposition that everyone loves himself and is clearly not equivalent to the original.

A wff in which all the quantifiers occur in an unbroken sequence at the beginning, with the scope of each extending to the end of the wff, is said to be in prenex normal form (PNF). Wffs that are in PNF are often more convenient to work with than those that are not. For every wff of LPC, however, there is an equivalent wff in PNF (often simply called its PNF). One effective method for finding the PNF of any given wff is the following:

Reletter bound variables as far as is necessary to ensure (a) that each quantifier contains a distinct variable and (b) that no variable in the wff occurs both bound and free.

Use definitions or PC equivalences to eliminate all operators except $\sim$, $\cdot$, and $\vee$.

Use the De Morgan laws and the rule of quantifier transformation to eliminate all occurrences of $\sim$ immediately before parentheses or quantifiers.

Gather all of the quantifiers into a sequence at the beginning in the order in which they appear in the wff and take the whole of what remains as their scope. Example:

$$(\forall x)\{[\phi x \cdot (\exists y)\psi xy] \supset (\exists y)\chi xy\} \supset (\exists z)(\phi z \supset \psi zx).$$

Step 1 can be achieved by relettering the third and fourth occurrences of y and every occurrence of x except the last (which is free); thus:

$$(\forall w)\{[\phi w \cdot (\exists y)\psi wy] \supset (\exists u)\chi wu\} \supset (\exists z)(\phi z \supset \psi zx).$$

Step 2 now yields:

$$\sim(\forall w)\{\sim[\phi w \cdot (\exists y)\psi wy] \vee (\exists u)\chi wu\} \vee (\exists z)(\sim\phi z \vee \psi zx).$$

By step 3 this becomes:

$$(\exists w)\{[\phi w \cdot (\exists y)\psi wy] \cdot (\forall u) \sim \chi wu\} \vee (\exists z)(\sim \phi z \vee \psi zx)$$

Finally, step 4 yields:

$$(\exists w)(\exists y)(\forall u)(\exists z)\{[(\phi w \cdot \psi wy) \cdot \sim \chi wu] \vee (\sim \phi z \vee \psi zx)\}$$

which is in PNF.

Classification of Dyadic Relations

Consider the closed wff:

$$(\forall x)(\forall y)(\phi xy \supset \phi yx)$$

which means that, whenever the relation ϕ holds between one object and a second, it also holds between that second object and the first. This expression is not valid, since it is true for some relations but false for others. A relation for which it is true is called a symmetrical relation (example: "is parallel to"). If the relation ϕ is such that, whenever it holds between one object and a second, it fails to hold between the second and the first—i.e., if ϕ is such that:

$$(\forall x)(\forall y)(\phi xy \supset \sim \phi yx)$$

Then ϕ is said to be asymmetrical (example: "is greater than"). A relation that is neither symmetrical nor asymmetrical is said to be nonsymmetrical. Thus, ϕ is nonsymmetrical if:

$$(\exists x)(\exists y)(\phi xy \cdot \phi yx) \cdot (\exists x)(\exists y)(\phi xy \cdot \sim \phi yx)$$

(Example: "loves").

Dyadic relations can also be characterized in terms of another threefold division: A relation ϕ is said to be transitive if, whenever it holds between one object and a second and also between that second object and a third, it holds between the first and the third—i.e., if:

$$(\forall x)(\forall y)(\forall z)[(\phi xy \cdot \phi yz) \supset \phi xz]$$

(Example: "is greater than"). An intransitive relation is one that, whenever it holds between one object and a second and also between that second and a third, fails to hold between the first and the third; i.e., ϕ is intransitive if:

$$(\forall x)(\forall y)(\forall z)[(\phi xy \cdot \phi yz) \supset \sim \phi xz]$$

(Example: "is father of"). A relation that is neither transitive nor intransitive is said to be nontransitive. Thus, ϕ is nontransitive if:

$$(\exists x)(\exists y)(\exists z)(\phi xy \cdot \phi yz \cdot \phi xz) \cdot (\exists x)(\exists y)(\exists z)(\phi xy \cdot \phi yz \cdot \sim \phi xz)$$

(Example: "is a first cousin of").

A relation ϕ that always holds between any object and itself is said to be reflexive; i.e., ϕ is reflexive if:

$$(\forall x)\phi xx$$

(Example: "is identical with"). If ϕ never holds between any object and itself—i.e., if:

$$\sim (\exists x)\phi xx$$

then ϕ is said to be irreflexive (example: "is greater than"). If ϕ is neither reflexive nor irreflexive—i.e., if:

$$(\exists x)\phi xx \cdot (\exists x) \sim \phi xx$$

Then ϕ is said to be nonreflexive (example: "admires").

A relation such as "is of the same length as" is not strictly reflexive, as some objects do not have a length at all and thus are not of the same length as anything, even themselves. But this relation is reflexive in the weaker sense that, whenever an object is of the same length as anything, it is of the same length as itself. Such a relation is said to be quasi-reflexive. Thus, ϕ is quasi-reflexive if:

$$(\forall x)[(\exists y)\phi xy \supset \phi xx].$$

A reflexive relation is of course also quasi-reflexive.

For the most part, these three classifications are independent of each other; thus a symmetrical relation may be transitive (like "is equal to") or intransitive (like "is perpendicular to") or nontransitive (like "is one mile distant from"). There are, however, certain limiting principles, of which the most important are:

- Every relation that is symmetrical and transitive is at least quasi-reflexive.
- Every asymmetrical relation is irreflexive.
- Every relation that is transitive and irreflexive is asymmetrical.

A relation that is reflexive, symmetrical, and transitive is called an equivalence relation.

Axiomatization of LPC

Rules of uniform substitution for predicate calculi, though formulable, are mostly very complicated, and, to avoid the necessity for these rules, axioms for these systems are therefore usually given by axiom schemata.

Axiom schemata:

- Any LPC substitution-instance of any valid wff of PC is an axiom.
- Any wff of the form $(\forall a)\alpha \supset \beta$ is an axiom if β is either identical with α or differs from it only in that, wherever α has a free occurrence of a, β has a free occurrence of some other individual variable b.
- Any wff of the form $(\forall a)(\alpha \supset \beta) \supset [\alpha \supset (\forall a)\beta]$ is an axiom, provided that α contains no free occurrence of a.

Transformation rules:

- Modus ponens.
- If α is a theorem, so is $(\forall a)\alpha$, where a is any individual variable (rule of universal generalization).

The axiom schemata call for some explanation and comment. By an LPC substitution-instance of a wff of PC is meant any result of uniformly replacing every propositional variable in that wff by a wff of LPC. Thus, one LPC substitution-instance of $(p \supset \sim q) \supset (q \supset \sim p)$ is $[\phi xy \supset \sim(\forall x)\psi x] \supset [(\forall x)\psi x \supset \sim\phi xy]$. Axiom schema 1 makes available in LPC all manipulations such as commutation, transposition, and distribution, which depend only on PC principles. Examples of wffs that are axioms by axiom schema 2 are $(\forall x)\phi x \supset \phi x$, $(\forall x)\phi x \supset \phi y$, and $(\forall x)(\exists y)\phi xy \supset (\exists y)\phi zy$. To see why it is necessary for the variable that replaces a to be free in β, consider the last example: Here a is x, α is $(\exists y)\phi xy$, in which x is free, and β is $(\exists y)\phi zy$, in which z is free and replaces x. But had y, which would become bound by the quantifier $(\exists y)$, been chosen as a replacement instead of z, the result would have been $(\forall x)(\exists y)\phi xy \supset (\exists y)\phi yy$, the invalidity of which can be seen intuitively by taking ϕxy to mean "x is a child of y," for then $(\forall x)(\exists y)\phi xy$ will mean that everyone is a child of someone, which is true, but $(\exists y)\phi yy$ will mean that someone is a child of himself, which is false. The need for the proviso in axiom schema 3 can also be seen from an example. Defiance of the proviso would give as an axiom $(\forall x)(\phi x \supset \psi x) \supset [\phi x \supset (\forall x)\psi x]$; if ϕx were taken to mean "x is a Spaniard," ψx to mean "x is a European," and the free occurrence of x (the first occurrence in the consequent) to stand for Francisco Franco, then the antecedent would mean that every Spaniard is a European, but the consequent would mean that, if Francisco Franco is a Spaniard, then everyone is a European.

It can be proved—though the proof is not an elementary one—that the theorems derivable from the above basis are precisely the wffs of LPC that are valid by the definition of validity on validity in LPC (see above Validity in LPC). Several other bases for LPC are known that also have this property. The axiom schemata and transformation rules here given are such that any purported proof of a theorem can be effectively checked to determine whether it really is a proof or not; nevertheless, theoremhood in LPC, like validity in LPC, is not effectively decidable, in that there is no effective method of telling with regard to any arbitrary wff whether it is a theorem or not. In this respect, axiomatic bases for LPC contrast with those for PC.

Semantic Tableaux

Since the 1980s another technique for determining the validity of arguments in either PC or LPC has gained some popularity, owing both to its ease of learning and to its straightforward implementation by computer programs. Originally suggested by the Dutch logician Evert W. Beth, it was more fully developed and publicized by the American mathematician and logician Raymond M. Smullyan. Resting on the observation that it is impossible for the premises of a valid argument to be true while the conclusion is false, this method attempts to interpret (or evaluate) the premises in such a way that they are all simultaneously satisfied and the negation of the conclusion is also satisfied. Success in such an effort would show the argument to be invalid, while failure to find such an interpretation would show it to be valid.

The construction of a semantic tableau proceeds as follows: express the premises and negation of the conclusion of an argument in PC using only negation (~) and disjunction (∨) as propositional connectives. Eliminate every occurrence of two negation signs in a sequence (e.g., ~~~~~a becomes ~a). Now construct a tree diagram branching downward such that each disjunction is replaced by two branches, one for the left disjunct and one for the right. The original disjunction is true if either branch is true. Reference to De Morgan's laws shows that a negation of a disjunction is true just in case the negations of both disjuncts are true [i.e., $\sim(p \vee q) \equiv (\sim p \cdot \sim q)$]. This semantic observation leads to the rule that the negation of a disjunction becomes one branch containing the negation of each disjunct:

$$\begin{array}{c} \sim(a \vee b) \\ | \\ \sim a \\ \sim b \end{array}$$

Consider the following argument:

$$\begin{array}{l} a \vee b \\ \underline{\sim a} \\ b \end{array}$$

Write:

$$\begin{array}{l} a \vee b \\ \sim a \\ \sim b \end{array}$$

Now strike out the disjunction and form two branches:

$$\begin{array}{ccc} & \sim a & \\ & \sim b & \\ \swarrow & & \searrow \\ a & & b \end{array}$$

Only if all the sentences in at least one branch are true is it possible for the original premises to be true and the conclusion false (equivalently for the negation of the conclusion). By tracing the line upward in each branch to the top of the tree, one observes that no valuation of a in the left branch will result in all the sentences in that branch receiving the value true (because of the presence of a and ~a). Similarly, in the right branch the presence of b and ~b makes it impossible for a valuation to result in all the sentences of the branch receiving the value true. These are all the possible branches; thus, it is impossible to find a situation in which the premises are true and the conclusion false. The original argument is therefore valid.

This technique can be extended to deal with other connectives:

$$\begin{array}{c} (a \cdot b) \\ | \\ a \\ b \end{array} \qquad \begin{array}{ccc} & \sim(a \cdot b) & \\ \swarrow & & \searrow \\ \sim a & & \sim b \end{array}$$

$$\begin{array}{ccc} & a \supset b & \\ \swarrow & & \searrow \\ \sim a & & b \end{array} \qquad \begin{array}{c} \sim(a \supset b) \\ | \\ a \\ \sim b \end{array}$$

Furthermore, in LPC, rules for instantiating quantified wffs need to be introduced. Clearly, any branch containing both $(\forall x)\phi x$ and $\sim\phi y$ is one in which not all the sentences in that branch can be simultaneously satisfied (under the assumption of ω-consistency). Again, if all the branches fail to be simultaneously satisfiable, the original argument is valid.

Special Systems of LPC

LPC as expounded above may be modified by either restricting or extending the range of wffs in various ways:

Partial systems of LPC. Some of the more important systems produced by restriction are here outlined:

- It may be required that every predicate variable be monadic while still allowing an infinite number of individual and predicate variables. The atomic wffs are then simply those consisting of a predicate variable followed by a single individual variable. Otherwise, the formation rules remain as before, and the definition of validity is also as before, though simplified in obvious ways. This system is known as the monadic LPC; it provides a logic of properties but not of relations. One important characteristic of this system is that it is decidable. (The introduction of even a single dyadic predicate variable, however, would make the system undecidable, and, in fact, even the system that contains only a single dyadic predicate variable and no other predicate variables at all has been shown to be undecidable.)

- A still simpler system can be formed by requiring (1) that every predicate variable be monadic, (2) that only a single individual variable (e.g., x) be used, (3) that every occurrence of this variable be bound, and (4) that no quantifier occur within the scope of any other. Examples of wffs of this system are (∀x)[ϕx ⊃ (ψx · χx)] ("Whatever is ϕ is both ψ and χ"); (∃x)(ϕx · ~ψx) ("There is something that is ϕ but not ψ"); and (∀x)(ϕx ⊃ ψx) ⊃ (∃x)(ϕx · ψx) ("If whatever is ϕ is ψ, then something is both ϕ and ψ"). The notation for this system can be simplified by omitting x everywhere and writing ∃ϕ for "Something is ϕ," ∀(ϕ ⊃ ψ) for "Whatever is ϕ is ψ," and so on. Although this system is more rudimentary even than the monadic LPC (of which it is a fragment), the forms of a wide range of inferences can be represented in it. It is also a decidable system, and decision procedures of an elementary kind can be given for it.

Extensions of LPC. More elaborate systems, in which a wider range of propositions can be expressed, have been constructed by adding to LPC new symbols of various types. The most straightforward of such additions are:

- One or more individual constants (say, a, b, ...): these constants are interpreted as names of specific individuals; formally they are distinguished from individual variables by the fact that they cannot occur within quantifiers; e.g., (∀x) is a quantifier but (∀a) is not.

- One or more predicate constants (say, A, B, ...), each of some specified degree, thought of as designating specific properties or relations.

A further possible addition, which calls for somewhat fuller explanation, consists of symbols designed to stand for functions. The notion of a function may be sufficiently explained for present purposes as follows. There is said to be a certain function of n arguments (or, of degree n) when there is a rule that specifies a unique object (called the value of the function) whenever all the arguments are specified. In the domain of human beings, for example, "the mother of —" is a monadic function (a function of one argument), since for every human being there is a unique individual who is his mother; and in the domain of the natural numbers (i.e., 0, 1, 2, ...), "the sum of — and —" is a function of two arguments, since for any pair of natural numbers there is a natural number that is their sum. A function symbol can be thought of as forming a name out of other names (its arguments); thus, whenever x and y name numbers, "the sum of x and y" also names a number, and similarly for other kinds of functions and arguments.

To enable functions to be expressed in LPC there may be added:

- One or more function variables (say, f, g, ...) or one or more function constants (say, F, G, ...) or both, each of some specified degree. The former are interpreted as ranging over functions of the degrees specified and the latter as designating specific functions of that degree.

When any or all of a–c are added to LPC, the formation rules listed in the first paragraph of the section on the lower predicate calculus need to be modified to enable the new symbols to be incorporated into wffs. This can be done as follows: A term is first defined as either (1) an individual variable or (2) an individual constant or (3) any expression formed by prefixing a function variable or function constant of degree n to any n terms (these terms—the arguments of the function symbol—are usually separated by commas and enclosed in parentheses). Formation rule 1 is then replaced by:

An expression consisting of a predicate variable or predicate constant of degree n followed by n terms is a wff.

The axiomatic basis also requires the following modification: in axiom schema 2 any term is allowed to replace a when β is formed, provided that no variable that is free in the term becomes bound in β. The following examples will illustrate the use of the aforementioned additions to LPC: let the values of the individual variables be the natural numbers; let the individual constants a and b stand for the numbers 2 and 3, respectively; let A mean "is prime"; and let F represent the dyadic function "the sum of." Then AF(a,b) expresses the proposition "The sum of 2 and 3 is prime," and (∃x) AF(x,a) expresses the proposition "There exists a number such that the sum of it and 2 is prime."

The introduction of constants is normally accompanied by the addition to the axiomatic basis of special axioms containing those constants, designed to express principles that hold of the objects, properties, relations, or functions represented by them—though they do not hold of objects, properties, relations, or functions in general. It may be decided, for example, to use the constant A to represent the dyadic relation "is greater than" (so that Axy is to mean "x is greater than y" and so forth). This relation, unlike many others, is transitive; i.e., if one object is greater than a second and that second is in turn greater than a third, then the first is greater than the third. Hence, the following special axiom schema might be added: if t_1, t_2, and t_3 are any terms, then:

$$(At_1t_2 \cdot At_2t_3) \supset At_1t_3$$

is an axiom. By such means systems can be constructed to express the logical structures of various particular disciplines. The area in which most work of this kind has been done is that of natural-number arithmetic.

PC and LPC are sometimes combined into a single system. This may be done most simply by adding propositional variables to the list of LPC primitives, adding a formation rule to the effect that a propositional variable standing alone is a wff, and deleting "LPC" in axiom schema 1. This yields as wffs such expressions as $(p \vee q) \supset (\forall x)\phi x$ and $(\exists x)[p \supset (\forall y)\phi xy]$.

LPC-with-identity. The word "is" is not always used in the same way. In a proposition such as (1) "Socrates is snub-nosed," the expression preceding the "is" names an individual and the expression following it stands for a property attributed to that individual. But, in a proposition such as (2) "Socrates is the Athenian philosopher who drank hemlock," the expressions preceding and following the "is" both name individuals, and the sense of the whole proposition is that the individual named by the first is the same individual as the individual named by the second. Thus, in 2 "is" can be expanded to "is the same individual as," whereas in 1 it cannot. As used in 2, "is" stands for a dyadic relation—namely, identity—that the proposition asserts to hold between the two individuals. An identity proposition is to be understood in this context as asserting no more than this; in particular it is not to be taken as asserting that the two naming expressions have the same meaning. A much-discussed example to illustrate this last point is "The morning star is the evening star." It is false that the expressions "the morning star" and "the evening star" mean the same, but it is true that the object referred to by the former is the same as that referred to by the latter (the planet Venus).

To enable the forms of identity propositions to be expressed, a dyadic predicate constant is added to LPC, for which the most usual notation is = (written between, rather than before, its arguments). The intended interpretation of $x = y$ is that x is the same individual as y, and the most convenient reading is "x is identical with y." Its negation $\sim(x = y)$ is commonly abbreviated as $x \neq y$. To the definition of an LPC model given earlier (see above Validity in LPC) there is now added the rule (which accords in an obvious way with the intended interpretation) that the value of $x = y$ is to be 1 if the same member of D is assigned to both x and y and that otherwise its value is to be 0; validity can then be defined as before. The following additions (or some equivalent ones) are made to the axiomatic basis for LPC: the axiom $x = x$ and the axiom schema that, where a and b are any individual variables and α and β are wffs that differ only in that, at one or more places where α has a free occurrence of a, β has a free occurrence of b, $(a = b) \supset (\alpha \supset \beta)$ is an axiom. Such a system is known as a lower-predicate-calculus-with-identity; it may of course be further augmented in the other ways referred to above in "Extensions of LPC," in which case any term may be an argument of =.

Identity is an equivalence relation; i.e., it is reflexive, symmetrical, and transitive. Its reflexivity is directly expressed in the axiom $x = x$, and theorems expressing its symmetry and transitivity can easily be derived from the basis given.

Certain wffs of LPC-with-identity express propositions about the number of things that possess a given property. "At least one thing is ϕ" could, of course, already be expressed by $(\exists x)\phi x$; "At least two distinct (nonidentical) things are ϕ" can now be expressed by $(\exists x)(\exists y)(\phi x \cdot \phi y \cdot x \neq y)$; and the sequence can be continued in an obvious way. "At most one thing is ϕ" (i.e., "No two distinct things are both ϕ") can be expressed by the negation of the last-mentioned wff or by its equivalent, $(\forall x)(\forall y)[(\phi x \cdot \phi y) \supset x = y]$, and the sequence can again be easily continued. A formula for "Exactly one thing is ϕ" may be obtained by conjoining the formulas for "At least one thing is ϕ" and "At most one thing is ϕ," but a simpler wff equivalent to this conjunction is $(\exists x)[\phi x \cdot (\forall y)(\phi y \supset x = y)]$, which means "There is something that is ϕ, and anything that is ϕ is that thing." The proposition "Exactly two things are ϕ" can be represented by:

$$(\exists x)(\exists y)\{\phi x \cdot \phi y \cdot x \neq y \cdot (\forall z)[\phi z \supset (z = x \vee z = y)]\}$$

i.e., "There are two nonidentical things each of which is ϕ, and anything that is ϕ is one or the other of these." Clearly, this sequence can also be extended to give a formula for "Exactly n things are ϕ" for every natural number n. It is convenient to abbreviate the wff for "Exactly one thing is ϕ" to $(\exists !x)\phi x$. This special quantifier is frequently read aloud as "E-Shriek x."

Definite Descriptions

When a certain property ϕ belongs to one and only one object, it is convenient to have an expression that names that object. A common notation for this purpose is $(\iota x)\phi x$, which may be read as "the thing that is ϕ" or more briefly as "the ϕ." In general, where a is any individual variable and α is any wff, $(\iota a)\alpha$ then stands for the single value of a that makes α true. An expression of the form "the so-and-so" is called a definite description; and (ιx), known as a description operator, can be thought of as forming a name of an individual out of a proposition form. (ιx) is analogous to a quantifier in that, when prefixed to a wff α, it binds every free occurrence of x in α. Relettering of bound variables is also permissible; in the simplest case, $(\iota x)\phi x$ and $(\iota y)\phi y$ can each be read simply as "the ϕ."

As far as formation rules are concerned, definite descriptions can be incorporated into LPC by letting expressions of the form $(\iota a)\alpha$ count as terms; rule 1′ above, in "Extensions of LPC," will then allow them to occur in atomic formulas (including identity formulas). "The ϕ is (i.e., has the property) ψ" can then be expressed as $\psi(\iota x)\phi x$; "y is (the same individual as) the ϕ" as $y = (\iota x)\phi x$; "The ϕ is (the same individual as) the ψ" as $(\iota x)\phi x = (\iota y)\psi y$; and so forth.

The correct analysis of propositions containing definite descriptions has been the subject of considerable philosophical controversy. One widely accepted account, however—substantially that presented in Principia Mathematica and known as Russell's theory of descriptions—holds that "The ϕ is ψ" is to be understood as meaning that exactly one thing is ϕ and that thing is also ψ. In that case it can be expressed by a wff of LPC-with-identity that contains no description operators—namely,

$$(\exists x)[\phi x \cdot (\forall y)(\phi y \supset x=y) \cdot \psi x]$$

Analogously, "y is the ϕ" is analyzed as "y is ϕ and nothing else is ϕ" and hence as expressible by:

$$\phi y \cdot (\forall x)(\phi x \supset x=y)$$

"The ϕ is the ψ" is analyzed as "Exactly one thing is ϕ, exactly one thing is ψ, and whatever is ϕ is ψ" and hence as expressible by:

$$(\exists x)[\phi x \cdot (\forall y)(\phi y \supset x=y)] \cdot (\exists x)[\psi x \cdot (\forall y)(\upsilon y \supset x=y)] \cdot (\forall x)(\phi x \supset \psi x)$$

$\psi(\iota x)\phi x$, $y = (\iota x)\phi x$ and $(\iota x)\phi x = (\iota y)\psi y$ can then be regarded as abbreviations and by generalizing to more complex cases, all wffs that contain description operators can be regarded as abbreviations for longer wffs that do not.

The analysis that leads to $(\exists x)[\phi x \cdot (\forall y)(\phi y \supset x=y) \cdot \psi x]$ as a formula for "The ϕ is ψ" leads to the following for "The ϕ is not ψ":

$$(\exists x)[\phi x \cdot (\forall y)(\phi y \supset x=y) \cdot \sim \psi x]$$

It is important to note that the equation above is not the negation of $(\exists x)[\phi x \cdot (\forall y)(\phi y \supset x=y) \cdot \psi x]$; this negation is, instead,

$$\sim (\exists x)[\phi x \cdot (\forall y)(\phi y \supset x=y) \cdot \psi x]$$

The difference in meaning between the above mentioned equation lies in the fact that it is true only when there is exactly one thing that is ϕ and that thing is not ψ, but it is true both in this case and also when nothing is ϕ at all and when more than one thing is ϕ. Neglect of the distinction between both above mentioned equations can result in serious confusion of thought; in ordinary speech it is frequently unclear whether someone who denies that the ϕ is ψ is conceding that exactly one thing is ϕ but denying that it is ψ, or denying that exactly one thing is ϕ.

The basic contention of Russell's theory of descriptions is that a proposition containing a definite description is not to be regarded as an assertion about an object of which that description is a name but rather as an existentially quantified assertion that a certain (rather complex) property has an instance. Formally, this is reflected in the rules for eliminating description operators that were outlined above.

Higher-order Predicate Calculi

A feature shared by LPC and all its extensions so far mentioned is that the only variables that occur in quantifiers are individual variables. It is by virtue of this feature that they are called lower (or first-order) calculi. Various predicate calculi of higher order can be formed, however, in which quantifiers may contain other variables as well, hence binding all free occurrences of these that lie within their scope. In particular, in the second-order predicate calculus, quantification is permitted over both individual and predicate variables; hence, wffs such as $(\forall\phi)(\exists x)\phi x$ can be formed. This last formula, since it contains no free variables of any kind, expresses a determinate proposition—namely, the proposition that every property has at least one instance. One important feature of this system is that in it identity need not be taken as primitive but can be introduced by defining $x = y$ as $(\forall\phi)(\phi x \equiv \phi y)$—i.e., "Every property possessed by x is also possessed by y and vice versa." Whether such a definition is acceptable as a general account of identity is a question that raises philosophical issues too complex to be discussed here; they are substantially those raised by the principle of the identity of indiscernibles, best known for its exposition in the 17th century by Gottfried Wilhelm Leibniz.

Modal Logic

True propositions can be divided into those—like "2 + 2 = 4"—that are true by logical necessity (necessary propositions), and those—like "France is a republic"—that are not (contingently true propositions). Similarly, false propositions can be divided into those—like "2 + 2 = 5"—that are false by logical necessity (impossible propositions), and those—like "France is a monarchy"—that are not (contingently false propositions). Contingently true and contingently false propositions are known collectively as contingent propositions. A proposition that is not impossible (i.e., one that is either necessary or contingent) is said to be a possible proposition. Intuitively, the notions of necessity and possibility are connected in the following way: to say that a proposition is necessary is to say that it is not possible for it to be false, and to say that a proposition is possible is to say that it is not necessarily false.

If it is logically impossible for a certain proposition, p, to be true without a certain proposition, q, being also true (i.e., if the conjunction of p and not-q is logically impossible), then it is said that p strictly implies q. An alternative equivalent way of explaining the notion of strict implication is by saying that p strictly implies q if and only if it is necessary that p materially implies q. "John's tie is scarlet," for example, strictly implies "John's tie is red," because it is impossible for John's tie to be scarlet without being red (or it is necessarily true that, if John's tie is scarlet, it is red). In general, if p is the conjunction of the premises, and q the conclusion, of a deductively valid inference, p will strictly imply q.

The notions just referred to—necessity, possibility, impossibility, contingency, strict implication—and certain other closely related ones are known as modal notions, and a logic designed to express principles involving them is called a modal logic.

The most straightforward way of constructing such a logic is to add to some standard nonmodal system a new primitive operator intended to represent one of the modal notions, to define other modal operators in terms of it, and to add certain special axioms or transformation rules or both. A great many systems of modal logic have been constructed, but attention will be restricted here to a few closely related ones in which the underlying nonmodal system is ordinary PC.

Alternative Systems of Modal Logic

All the systems to be considered here have the same wffs but differ in their axioms. The wffs can be specified by adding to the symbols of PC a primitive monadic operator L and to the formation rules of PC the rule that if α is a wff, so is Lα. L is intended to be interpreted as "It is necessary that," so that Lp will be true if and only if p is a necessary proposition. The monadic operator M and the dyadic operator Z (to be interpreted as "It is possible that" and "strictly implies," respectively) can then be introduced by the following definitions, which reflect in an obvious way the informal accounts given above of the connections between necessity, possibility, and strict implication: if α is any wff, then Mα is to be an abbreviation of $\sim L \sim \alpha$; and if α and β are any wffs, then α Z β is to be an abbreviation of $L(\alpha \supset \beta)$ [or alternatively of $\sim M(\alpha \cdot \sim\beta)$].

The modal system known as T has as axioms some set of axioms adequate for PC (such as those of PM), and in addition:

- $Lp \supset p$.
- $L(p \supset q) \supset (Lp \supset Lq)$.

Axiom 1 expresses the principle that whatever is necessarily true is true, and 2 the principle that, if q logically follows from p, then, if p is a necessary truth, so is q (i.e., that whatever follows from a necessary truth is itself a necessary truth). These two principles seem to have a high degree of intuitive plausibility, and 1 and 2 are theorems in almost all modal systems. The transformation rules of T are uniform substitution, modus ponens, and a rule to the effect that if α is a theorem so is Lα (the rule of necessitation). The intuitive rationale of this rule is that, in a sound axiomatic system, it is expected that every instance of a theorem α will be not merely true but necessarily true—and in that case every instance of Lα will be true.

Among the simpler theorems of T are:

- $p \supset Mp$.
- $L(p \cdot q) \equiv (Lp \cdot Lq)$.
- $M(p \vee q) \equiv (Mp \vee Mq)$.
- $(Lp \vee Lq) \supset L(p \vee q)$ (but not its converse).
- $M(p \cdot q) \supset (Mp \cdot Mq)$ (but not its converse).

and,

- LMp ≡ ~ML~p.
- (p Z q) ⊃ (Mp ⊃ Mq).
- (~p Z p) ≡ Lp.
- L(p ∨ q) ⊃ (Lp ∨ Mq).

There are many modal formulas that are not theorems of T but that have a certain claim to express truths about necessity and possibility. Among them are:

Lp ⊃ LLp, Mp ⊃ LMp, and p ⊃ LMp

The first of these means that if a proposition is necessary, its being necessary is itself a necessary truth; the second means that if a proposition is possible, its being possible is a necessary truth; and the third means that if a proposition is true, then not merely is it possible but its being possible is a necessary truth. These are all various elements in the general thesis that a proposition's having the modal characteristics it has (such as necessity, possibility) is not a contingent matter but is determined by logical considerations. Although this thesis may be philosophically controversial, it is at least plausible, and its consequences are worth exploring. One way of exploring them is to construct modal systems in which the formulas listed above are theorems. None of these formulas, as was said, is a theorem of T; but each could be consistently added to T as an extra axiom to produce a new and more extensive system. The system obtained by adding Lp ⊃ LLp to T is known as S4; that obtained by adding Mp ⊃ LMp to T is known as S5; and the addition of p ⊃ LMp to T gives the Brouwerian system, here called B for short.

The relations between these four systems are as follows: S4 is stronger than T; i.e., it contains all the theorems of T and others besides. B is also stronger than T. S5 is stronger than S4 and also stronger than B. S4 and B, however, are independent of each other in the sense that each contains some theorems that the other does not have. It is of particular importance that, if Mp ⊃ LMp is added to T, then Lp ⊃ LLp can be derived as a theorem, but, if one merely adds the latter to T, the former cannot then be derived.

Examples of theorems of S4 that are not theorems of T are Mp ≡ MMp, MLMp ⊃ Mp, and (p Z q) ⊃ (Lp Z Lq). Examples of theorems of S5 that are not theorems of S4 are Lp ≡ MLp, L(p ∨ Mq) ≡ (Lp ∨ Mq), M(p · Lq) ≡ (Mp · Lq), and (Lp Z Lq) ∨ (Lq Z Lp). One important feature of S5 but not of the other systems mentioned is that any wff that contains an unbroken sequence of monadic modal operators (Ls or Ms or both) is probably equivalent to the same wff with all these operators deleted except the last.

Considerations of space preclude an account of the many other axiomatic systems of modal logic that have been investigated. Some of these are weaker than T; such systems normally contain the axioms of T either as axioms or as theorems but have only a restricted form of the rule of necessitation. Another group comprises systems that are stronger than S4 but weaker than S5; some of these have proved fruitful in developing a logic of temporal relations. Yet another group includes systems that are stronger than S4 but independent of S5 in the sense explained above.

Modal predicate logics can also be formed by making analogous additions to LPC instead of to PC.

Validity in Modal Logic

The task of defining validity for modal wffs is complicated by the fact that, even if the truth values of all of the variables in a wff are given, it is not obvious how one should set about calculating the truth value of the whole wff. Nevertheless, a number of definitions of validity applicable to modal wffs have been given, each of which turns out to match some axiomatic modal system in the sense that it brings out as valid those wffs, and no others, that are theorems of that system. Most, if not all, of these accounts of validity can be thought of as variant ways of giving formal precision to the idea that necessity is truth in every possible world or conceivable state of affairs. The simplest such definition is this: let a model be constructed by first assuming a (finite or infinite) set W of worlds. In each world, independently of all the others, let each propositional variable then be assigned either the value 1 or the value 0. In each world the values of truth functions are calculated in the usual way from the values of their arguments in that world. In each world, however, Lα is to have the value 1 if α has the value 1 not only in that world but in every other world in W as well and is otherwise to have the value 0; and in each world Mα is to have the value 1 if α has value 1 either in that world or in some other world in W and is otherwise to have the value 0. These rules enable one to calculate a value (1 or 0) in any world in W for any given wff, once the values of the variables in each world in W are specified. A model is defined as consisting of a set of worlds together with a value assignment of the kind just described. A wff is valid if and only if it has the value 1 in every world in every model. It can be proved that the wffs that are valid by this criterion are precisely the theorems of S5; for this reason models of the kind here described may be called S5-models, and validity as just defined may be called S5-validity.

A definition of T-validity (i.e., one that can be proved to bring out as valid precisely the theorems of T) can be given as follows: a T-model consists of a set of worlds W and a value assignment to each variable in each world, as before. It also includes a specification, for each world in W, of some subset of W as the worlds that are "accessible" to that world. Truth functions are evaluated as before, but, in each world in the model, Lα is to have the value 1 if α has the value 1 in that world and in every other world in W accessible to it and is otherwise to have the value 0. And, in each world, Mα is to have the value 1 if α has the value 1 either in that world or in some other world accessible to it and is otherwise to have the value 0. (In other words, in computing the value of

Lα or Mα in a given world, no account is taken of the value of α in any other world not accessible to it.) A wff is T-valid if and only if it has the value 1 in every world in every T-model.

An S4-model is defined as a T-model except that it is required that the accessibility relation be transitive—i.e., that, where w1, w2, and w3 are any worlds in W, if w1 is accessible to w2 and w2 is accessible to w3, then w1 is accessible to w3. A wff is S4-valid if and only if it has the value 1 in every world in every S4-model. The S4-valid wffs can be shown to be precisely the theorems of S4. Finally, a definition of validity is obtained that will match the system B by requiring that the accessibility relation be symmetrical but not that it be transitive.

For all four systems, effective decision procedures for validity can be given. Further modifications of the general method described have yielded validity definitions that match many other axiomatic modal systems, and the method can be adapted to give a definition of validity for intuitionistic PC. For a number of axiomatic modal systems, however, no satisfactory account of validity has been devised. Validity can also be defined for various modal predicate logics by combining the definition of LPC-validity given earlier with the relevant accounts of validity for modal systems, but a modal logic based on LPC is, like LPC itself, an undecidable system.

First-order Logic

First-order logic—also known as predicate logic, quantificational logic, and first-order predicate calculus—is a collection of formal systems used in mathematics, philosophy, linguistics, and computer science. First-order logic uses quantified variables over non-logical objects and allows the use of sentences that contain variables, so that rather than propositions such as *Socrates is a man* one can have expressions in the form “there exists x such that x is Socrates and x is a man” and *there exists* is a quantifier while *x* is a variable. This distinguishes it from propositional logic, which does not use quantifiers or relations; in this sense, propositional logic is the foundation of first-order logic.

A theory about a topic is usually a first-order logic together with a specified domain of discourse over which the quantified variables range, finitely many functions from that domain to itself, finitely many predicates defined on that domain, and a set of axioms believed to hold for those things. Sometimes “theory” is understood in a more formal sense, which is just a set of sentences in first-order logic.

The adjective “first-order” distinguishes first-order logic from higher-order logic in which there are predicates having predicates or functions as arguments, or in which one or both of predicate quantifiers or function quantifiers are permitted. In first-order theories, predicates are often associated with sets. In interpreted higher-order theories, predicates may be interpreted as sets of sets.

There are many deductive systems for first-order logic which are both sound (all provable statements are true in all models) and complete (all statements which are true in all models are provable). Although the logical consequence relation is only semidecidable, much progress has been made in automated theorem proving in first-order logic. First-order logic also satisfies several metalogical theorems that make it amenable to analysis in proof theory, such as the Löwenheim–Skolem theorem and the compactness theorem.

First-order logic is the standard for the formalization of mathematics into axioms and is studied in the foundations of mathematics. Peano arithmetic and Zermelo–Fraenkel set theory are axiomatizations of number theory and set theory, respectively, into first-order logic. No first-order theory, however, has the strength to uniquely describe a structure with an infinite domain, such as the natural numbers or the real line. Axiom systems that do fully describe these two structures (that is, categorical axiom systems) can be obtained in stronger logics such as second-order logic.

The foundations of first-order logic were developed independently by Gottlob Frege and Charles Sanders Peirce. For a history of first-order logic and how it came to dominate formal logic.

While propositional logic deals with simple declarative propositions, first-order logic additionally covers predicates and quantification.

A predicate takes an entity or entities in the domain of discourse as input while outputs are either True or False. Consider the two sentences "Socrates is a philosopher" and "Plato is a philosopher". In propositional logic, these sentences are viewed as being unrelated and might be denoted, for example, by variables such as p and q. The predicate "is a philosopher" occurs in both sentences, which have a common structure of "a is a philosopher". The variable a is instantiated as "Socrates" in the first sentence and is instantiated as "Plato" in the second sentence. While first-order logic allows for the use of predicates, such as "is a philosopher" in this example, propositional logic does not.

Relationships between predicates can be stated using logical connectives. Consider, for example, the first-order formula "if a is a philosopher, then a is a scholar". This formula is a conditional statement with "a is a philosopher" as its hypothesis and "a is a scholar" as its conclusion. The truth of this formula depends on which object is denoted by a, and on the interpretations of the predicates "is a philosopher" and "is a scholar".

Quantifiers can be applied to variables in a formula. The variable a in the previous formula can be universally quantified, for instance, with the first-order sentence "For every a, if a is a philosopher, then a is a scholar". The universal quantifier "for every" in this sentence expresses the idea that the claim "if a is a philosopher, then a is a scholar" holds for all choices of a.

The negation of the sentence "For every a, if a is a philosopher, then a is a scholar" is logically equivalent to the sentence "There exists a such that a is a philosopher and a is not a scholar". The existential quantifier "there exists" expresses the idea that the claim "a is a philosopher and a is not a scholar" holds for some choice of a.

The predicates "is a philosopher" and "is a scholar" each take a single variable. In general, predicates can take several variables. In the first-order sentence "Socrates is the teacher of Plato", the predicate "is the teacher of" takes two variables.

An interpretation (or model) of a first-order formula specifies what each predicate means and the entities that can instantiate the variables. These entities form the domain of discourse or universe, which is usually required to be a nonempty set. For example, in an interpretation with the domain of discourse consisting of all human beings and the predicate "is a philosopher" understood as "was the author of the Republic", the sentence "There exists a such that a is a philosopher" is seen as being true, as witnessed by Plato.

Syntax

There are two key parts of first-order logic. The syntax determines which finite sequences of symbols are legal expressions in first-order logic, while the semantics determine the meanings behind these expressions.

Alphabet

Unlike natural languages, such as English, the language of first-order logic is completely formal, so that it can be mechanically determined whether a given expression is legal. There are two key types of legal expressions: terms, which intuitively represent objects, and formulas, which intuitively express predicates that can be true or false. The terms and formulas of first-order logic are strings of symbols, where all the symbols together form the alphabet of the language. As with all formal languages, the nature of the symbols themselves is outside the scope of formal logic; they are often regarded simply as letters and punctuation symbols.

It is common to divide the symbols of the alphabet into logical symbols, which always have the same meaning, and non-logical symbols, whose meaning varies by interpretation. For example, the logical symbol $\wedge$ always represents "and"; it is never interpreted as "or". On the other hand, a non-logical predicate symbol such as Phil(x) could be interpreted to mean "x is a philosopher", "x is a man named Philip", or any other unary predicate, depending on the interpretation at hand.

Logical Symbols

There are several logical symbols in the alphabet, which vary by author but usually include:

- The quantifier symbols ∀ and ∃.
- The logical connectives: ∧ for conjunction, ∨ for disjunction, → for implication, ↔ for biconditional, ¬ for negation. Occasionally other logical connective symbols are included. Some authors use C*pq*, instead of →, and E*pq*, instead of ↔, especially in contexts where → is used for other purposes. Moreover, the horseshoe ⊃ may replace →; the triple-bar ≡ may replace ↔; a tilde (~), N*p*, or F*pq*, may replace ¬; ||, or A*pq* may replace ∨; and and, K*pq*, or the middle dot, ·, may replace ∧, especially if these symbols are not available for technical reasons. (*Note*: the aforementioned symbols C*pq*, E*pq*, N*p*, A*pq*, and K*pq* are used in Polish notation.)
- Parentheses, brackets, and other punctuation symbols. The choice of such symbols varies depending on context.
- An infinite set of variables, often denoted by lowercase letters at the end of the alphabet x, y, z, … . Subscripts are often used to distinguish variables: x_0, x_1, x_2, … .
- An equality symbol (sometimes, identity symbol) =.

Not all of these symbols are required–only one of the quantifiers, negation and conjunction, variables, brackets and equality suffice. There are numerous minor variations that may define additional logical symbols:

- Sometimes the truth constants T, V*pq*, or ⊤, for "true" and F, O*pq*, or ⊥, for "false" are included. Without any such logical operators of valence 0, these two constants can only be expressed using quantifiers.
- Sometimes additional logical connectives are included, such as the Sheffer stroke, D*pq* (NAND), and exclusive or, J*pq*.

Non-logical Symbols

The non-logical symbols represent predicates (relations), functions and constants on the domain of discourse. It used to be standard practice to use a fixed, infinite set of non-logical symbols for all purposes. A more recent practice is to use different non-logical symbols according to the application one has in mind. Therefore, it has become necessary to name the set of all non-logical symbols used in a particular application. This choice is made via a signature.

The traditional approach is to have only one, infinite, set of non-logical symbols (one signature) for all applications. Consequently, under the traditional approach there is only one language of first-order logic.

- For every integer n ≥ 0 there is a collection of n-ary, or n-place, predicate symbols. Because they represent relations between n elements, they are also called relation symbols. For each arity n we have an infinite supply of them:

  $P^n_0, P^n_1, P^n_2, P^n_3, \ldots$

- For every integer n ≥ 0 there are infinitely many n-ary function symbols:

  $f^n_0, f^n_1, f^n_2, f^n_3, \ldots$

In contemporary mathematical logic, the signature varies by application. Typical signatures in mathematics are {1, ×} or just {×} for groups, or {0, 1, +, ×, <} for ordered fields. There are no restrictions on the number of non-logical symbols. The signature can be empty, finite, or infinite, even uncountable. Uncountable signatures occur for example in modern proofs of the Löwenheim–Skolem theorem.

In this approach, every non-logical symbol is of one of the following types.

- A predicate symbol (or relation symbol) with some valence (or arity, number of arguments) greater than or equal to 0. These are often denoted by uppercase letters P, Q, R,... .
  - Relations of valence 0 can be identified with propositional variables. For example, P, which can stand for any statement.
  - For example, P(x) is a predicate variable of valence 1. One possible interpretation is "x is a man".
  - Q(x,y) is a predicate variable of valence 2. Possible interpretations include "x is greater than y" and "x is the father of y".
- A function symbol, with some valence greater than or equal to 0. These are often denoted by lowercase letters f, g, h,... .
  - Examples: f(x) may be interpreted as for "the father of x". In arithmetic, it may stand for "-x". In set theory, it may stand for "the power set of x". In arithmetic, g(x,y) may stand for "x+y". In set theory, it may stand for "the union of x and y".
  - Function symbols of valence 0 are called constant symbols, and are often denoted by lowercase letters at the beginning of the alphabet a, b, c,... . The symbol a may stand for Socrates. In arithmetic, it may stand for 0. In set theory, such a constant may stand for the empty set.

The traditional approach can be recovered in the modern approach by simply specifying the "custom" signature to consist of the traditional sequences of non-logical symbols.

Formation Rules

The formation rules define the terms and formulas of first-order logic. When terms and formulas are represented as strings of symbols, these rules can be used to write a formal grammar for terms and formulas. These rules are generally context-free (each production has a single symbol on the left side), except that the set of symbols may be allowed to be infinite and there may be many start symbols, for example the variables in the case of terms.

Terms

The set of terms is inductively defined by the following rules:

- Variables: Any variable is a term.
- Functions: Any expression $f(t_1,...,t_n)$ of n arguments (where each argument t_i is a term and f is a function symbol of valence n) is a term. In particular, symbols denoting individual constants are nullary function symbols, and are thus terms.

Only expressions which can be obtained by finitely many applications of rules 1 and 2 are terms. For example, no expression involving a predicate symbol is a term.

Formulas

The set of formulas (also called well-formed formulas or WFFs) is inductively defined by the following rules:

- Predicate symbols: If P is an n-ary predicate symbol and $t_1, ..., t_n$ are terms then $P(t_1,...,t_n)$ is a formula.
- Equality: If the equality symbol is considered part of logic, and t_1 and t_2 are terms, then $t_1 = t_2$ is a formula.
- Negation: If φ is a formula, then $\neg\varphi$ is a formula.
- Binary connectives: If φ and ψ are formulas, then $(\varphi\rightarrow\psi)$ is a formula. Similar rules apply to other binary logical connectives.
- Quantifiers: If φ is a formula and x is a variable, then $\forall x\varphi$ (for all x, φ holds) and $\exists x\varphi$ (there exists x such that φ) are formulas.

Only expressions which can be obtained by finitely many applications of rules 1–5 are formulas. The formulas obtained from the first two rules are said to be atomic formulas.

For example,

$$\forall x \forall y \left(P\left(f\left(x\right)\right)\right) \rightarrow \neg Q\left(f\left(y\right), x, z\right)$$

is a formula, if *f* is a unary function symbol, *P* a unary predicate symbol, and Q a ternary predicate symbol. On the other hand, $\forall x x \rightarrow$ is not a formula, although it is a string of symbols from the alphabet.

The role of the parentheses in the definition is to ensure that any formula can only be obtained in one way by following the inductive definition (in other words, there is a unique parse tree for each formula). This property is known as unique readability of formulas. There are many conventions for where parentheses are used in formulas. For example, some authors use colons or full stops instead of parentheses, or change the places in which parentheses are inserted. Each author›s particular definition must be accompanied by a proof of unique readability.

This definition of a formula does not support defining an if-then-else function ite(c, a, b), where "c" is a condition expressed as a formula, that would return "a" if c is true, and "b" if it is false. This is because both predicates and functions can only accept terms as parameters, but the first parameter is a formula. Some languages built on first-order logic, such as SMT-LIB 2.0, add this.

Notational Conventions

For convenience, conventions have been developed about the precedence of the logical operators, to avoid the need to write parentheses in some cases. These rules are similar to the order of operations in arithmetic. A common convention is:

- $\neg$ is evaluated first.
- $\wedge$ and $\vee$ are evaluated next.
- Quantifiers are evaluated next.
- $\vee$ is evaluated last.

Moreover, extra punctuation not required by the definition may be inserted to make formulas easier to read. Thus the formula:

$$\neg \forall x P\left(x\right) \rightarrow \exists x \neg P\left(x\right)$$

might be written as:

$$\left(\neg\left[\forall x P\left(x\right)\right]\right) \rightarrow \exists x\left[\neg P\left(x\right)\right].$$

In some fields, it is common to use infix notation for binary relations and functions, instead of the prefix notation defined above. For example, in arithmetic, one typically writes "2 + 2 = 4" instead of "=(+(2,2),4)". It is common to regard formulas in infix notation as abbreviations for the corresponding formulas in prefix notation, cf. also term structure vs. representation.

The definitions above use infix notation for binary connectives such as $\rightarrow$. A less common convention is Polish notation, in which one writes $\rightarrow$, $\wedge$, and so on in front of their arguments rather than between them. This convention allows all punctuation symbols to be discarded. Polish notation is compact and elegant, but rarely used in practice because it is hard for humans to read it. In Polish notation, the formula:

$$\forall x \forall y \left(P\left(f\left(x\right)\right) \rightarrow \neg \left(P\left(x\right) \rightarrow Q\left(f\left(y\right), x, z\right)\right)\right)$$

becomes "$\forall x \forall y \rightarrow Pfx \neg \rightarrow PxQfyxz$".

Free and Bound Variables

In a formula, a variable may occur free or bound (or both). Intuitively, a variable occurrence is free in a formula if it is not quantified: in $\forall y P(x,y)$, the sole occurrence of variable x is free while that of y is bound. The free and bound variable occurrences in a formula are defined inductively as follows.

- Atomic formulas: If φ is an atomic formula then x occurs free in φ if and only if x occurs in φ. Moreover, there are no bound variables in any atomic formula.
- Negation: x occurs free in $\neg\varphi$ if and only if x occurs free in φ. x occurs bound in $\neg\varphi$ if and only if x occurs bound in φ.
- Binary connectives: x occurs free in $(\varphi \rightarrow \psi)$ if and only if x occurs free in either φ or ψ. x occurs bound in $(\varphi \rightarrow \psi)$ if and only if x occurs bound in either φ or ψ. The same rule applies to any other binary connective in place of $\rightarrow$.
- Quantifiers: x occurs free in $\forall y \varphi$ if and only if x occurs free in φ and x is a different symbol from y. Also, x occurs bound in $\forall y \varphi$ if and only if x is y or x occurs bound in φ. The same rule holds with $\exists$ in place of $\forall$.

For example, in $\forall x \forall y\ (P(x) \rightarrow Q(x, f(x), z))$, x and y occur only bound, z occurs only free, and w is neither because it does not occur in the formula.

Free and bound variables of a formula need not be disjoint sets: in the formula $P(x) \rightarrow \forall x Q(x)$, the first occurrence of x, as argument of P, is free while the second one, as argument of Q, is bound.

A formula in first-order logic with no free variable occurrences is called a first-order sentence. These are the formulas that will have well-defined truth values under an interpretation. For example, whether a formula such as Phil(x) is true must depend on what x represents. But the sentence $\exists x\, \text{Phil}(x)$ will be either true or false in a given interpretation.

Example: Ordered Abelian Groups.

In mathematics the language of ordered abelian groups has one constant symbol 0, one unary function symbol −, one binary function symbol +, and one binary relation symbol ≤. Then:

- The expressions +(x, y) and +(x, +(y, −(z))) are terms. These are usually written as x + y and x + y − z.
- The expressions +(x, y) = 0 and ≤(+(x, +(y, −(z))), +(x, y)) are atomic formulas. These are usually written as x + y = 0 and x + y − z ≤ x + y.
- The expression $(\forall x \forall y[\leq(+(x,y),z) \rightarrow \forall x \forall y +(x,y)=0)]$ is a formula, which is usually written as $\forall x \forall y(x+y \leq z) \rightarrow \forall x \forall y(x+y=0)$. This formula has one free variable, z.

The axioms for ordered abelian groups can be expressed as a set of sentences in the language. For example, the axiom stating that the group is commutative is usually written $(\forall x)(\forall y)[x+y=y+x]$.

Semantics

An interpretation of a first-order language assigns a denotation to each non-logical symbol in that language. It also determines a domain of discourse that specifies the range of the quantifiers. The result is that each term is assigned an object that it represents, each predicate is assigned a property of objects, and each sentence is assigned a truth value. In this way, an interpretation provides semantic meaning to the terms, the predicates, and formulas of the language. The study of the interpretations of formal languages is called formal semantics. What follows is a description of the standard or Tarskian semantics for first-order logic. (It is also possible to define game semantics for first-order logic, but aside from requiring the axiom of choice, game semantics agree with Tarskian semantics for first-order logic, so game semantics will not be elaborated herein.)

The domain of discourse D is a nonempty set of "objects" of some kind. Intuitively, a first-order formula is a statement about these objects; for example, $\exists x P(x)$ states the existence of an object x such that the predicate P is true where referred to it. The domain of discourse is the set of considered objects. For example, one can take D to be the set of integer numbers.

The interpretation of a function symbol is a function. For example, if the domain of discourse consists of integers, a function symbol f of arity 2 can be interpreted as the function that gives the sum of its arguments. In other words, the symbol f is associated with the function I(f) which, in this interpretation, is addition.

The interpretation of a constant symbol is a function from the one-element set D^0 to D, which can be simply identified with an object in D. For example, an interpretation may assign the value $I(c) = 10$ to the constant symbol c.

The interpretation of an n-ary predicate symbol is a set of n-tuples of elements of the domain of discourse. This means that, given an interpretation, a predicate symbol, and n elements of the domain of discourse, one can tell whether the predicate is true of those elements according to the given interpretation. For example, an interpretation I(P) of a binary predicate symbol P may be the set of pairs of integers such that the first one is less than the second. According to this interpretation, the predicate *P* would be true if its first argument is less than the second.

First-order Structures

The most common way of specifying an interpretation (especially in mathematics) is to specify a structure (also called a model). The structure consists of a nonempty set D that forms the domain of discourse and an interpretation I of the non-logical terms of the signature. This interpretation is itself a function:

- Each function symbol f of arity n is assigned a function I(f) from D^n to D. In particular, each constant symbol of the signature is assigned an individual in the domain of discourse.
- Each predicate symbol P of arity n is assigned a relation I(P) over D^n or, equivalently, a function from D^n to $\{\text{true}, \text{false}\}$. Thus each predicate symbol is interpreted by a Boolean-valued function on D.

Evaluation of Truth Values

A formula evaluates to true or false given an interpretation, and a variable assignment μ that associates an element of the domain of discourse with each variable. The reason that a variable assignment is required is to give meanings to formulas with free variables, such as $y = x$. The truth value of this formula changes depending on whether *x* and *y* denote the same individual.

First, the variable assignment μ can be extended to all terms of the language, with the result that each term maps to a single element of the domain of discourse. The following rules are used to make this assignment:

- Variables: Each variable x evaluates to μ(x).
- Functions: Given terms $t_1, \ldots, t_n$ that have been evaluated to elements $d_1, \ldots, d_n$

of the domain of discourse, and a n-ary function symbol f, the term $f(t_1,\ldots,t_n)$ evaluates to $(I(f))(d_1,\ldots,d_n)$.

Next, each formula is assigned a truth value. The inductive definition used to make this assignment is called the T-schema.

- Atomic formulas (1): A formula $P(t_1,\ldots,t_n)$ is associated the value true or false depending on whether $\langle v_1,\ldots,v_n\rangle \in I(P)$, where $v_1,\ldots,v_n$ are the evaluation of the terms $t_1,\ldots,t_n$ and $I(P)$ is the interpretation of P, which by assumption is a subset of D^n.
- Atomic formulas (2): A formula $t_1 = t_2$ is assigned true if t_1 and t_2 evaluate to the same object of the domain of discourse.
- Logical connectives: A formula in the form $\neg\phi$, $\phi \rightarrow \psi$, etc. is evaluated according to the truth table for the connective in question, as in propositional logic.
- Existential quantifiers: A formula $\exists x\phi(x)$ is true according to M and μ if there exists an evaluation μ' of the variables that only differs from μ regarding the evaluation of x and such that φ is true according to the interpretation M and the variable assignment μ'. This formal definition captures the idea that $\exists x\phi(x)$ is true if and only if there is a way to choose a value for x such that $\varphi(x)$ is satisfied.
- Universal quantifiers: A formula $\forall x\phi(x)$ is true according to M and μ if $\varphi(x)$ is true for every pair composed by the interpretation M and some variable assignment μ' that differs from μ only on the value of x. This captures the idea that $\forall x\phi(x)$ is true if every possible choice of a value for x causes $\varphi(x)$ to be true.

If a formula does not contain free variables, and so is a sentence, then the initial variable assignment does not affect its truth value. In other words, a sentence is true according to M and μ if and only if it is true according to M and every other variable assignment μ'.

There is a second common approach to defining truth values that does not rely on variable assignment functions. Instead, given an interpretation M, one first adds to the signature a collection of constant symbols, one for each element of the domain of discourse in M; say that for each d in the domain the constant symbol c_d is fixed. The interpretation is extended so that each new constant symbol is assigned to its corresponding element of the domain. One now defines truth for quantified formulas syntactically, as follows:

- Existential quantifiers (alternate): A formula $\exists x\phi(x)$ is true according to M if there is some d in the domain of discourse such that $\phi(c_d)$ holds. Here $\phi(c_d)$ is the result of substituting c_d for every free occurrence of x in φ.

- Universal quantifiers (alternate): A formula $\forall x\phi(x)$ is true according to M if, for every d in the domain of discourse, $\phi(c_d)$ is true according to M.

This alternate approach gives exactly the same truth values to all sentences as the approach via variable assignments.

Validity, Satisfiability and Logical Consequence

If a sentence φ evaluates to True under a given interpretation *M*, one says that *M* satisfies φ; this is denoted $M \vDash \varphi$. A sentence is satisfiable if there is some interpretation under which it is true.

Satisfiability of formulas with free variables is more complicated, because an interpretation on its own does not determine the truth value of such a formula. The most common convention is that a formula with free variables is said to be satisfied by an interpretation if the formula remains true regardless which individuals from the domain of discourse are assigned to its free variables. This has the same effect as saying that a formula is satisfied if and only if its universal closure is satisfied.

A formula is logically valid (or simply valid) if it is true in every interpretation. These formulas play a role similar to tautologies in propositional logic.

A formula φ is a logical consequence of a formula ψ if every interpretation that makes ψ true also makes φ true. In this case one says that φ is logically implied by ψ.

Algebraizations

An alternate approach to the semantics of first-order logic proceeds via abstract algebra. This approach generalizes the Lindenbaum–Tarski algebras of propositional logic. There are three ways of eliminating quantified variables from first-order logic that do not involve replacing quantifiers with other variable binding term operators:

- Cylindric algebra, by Alfred Tarski and colleagues.
- Polyadic algebra, by Paul Halmos.
- Predicate functor logic, mainly due to Willard Quine.

These algebras are all lattices that properly extend the two-element Boolean algebra.

Tarski and Givant showed that the fragment of first-order logic that has no atomic sentence lying in the scope of more than three quantifiers has the same expressive power as relation algebra. This fragment is of great interest because it suffices for Peano arithmetic and most axiomatic set theory, including the canonical ZFC. They also prove that first-order logic with a primitive ordered pair is equivalent to a relation algebra with two ordered pair projection functions.

First-order Theories, Models and Elementary Classes

A first-order theory of a particular signature is a set of axioms, which are sentences consisting of symbols from that signature. The set of axioms is often finite or recursively enumerable, in which case the theory is called effective. Some authors require theories to also include all logical consequences of the axioms. The axioms are considered to hold within the theory and from them other sentences that hold within the theory can be derived.

A first-order structure that satisfies all sentences in a given theory is said to be a model of the theory. An elementary class is the set of all structures satisfying a particular theory. These classes are a main subject of study in model theory.

Many theories have an intended interpretation, a certain model that is kept in mind when studying the theory. For example, the intended interpretation of Peano arithmetic consists of the usual natural numbers with their usual operations. However, the Löwenheim–Skolem theorem shows that most first-order theories will also have other, nonstandard models.

A theory is consistent if it is not possible to prove a contradiction from the axioms of the theory. A theory is complete if, for every formula in its signature, either that formula or its negation is a logical consequence of the axioms of the theory. Gödel's incompleteness theorem shows that effective first-order theories that include a sufficient portion of the theory of the natural numbers can never be both consistent and complete.

Empty Domains

The definition above requires that the domain of discourse of any interpretation must be nonempty. There are settings, such as inclusive logic, where empty domains are permitted. Moreover, if a class of algebraic structures include an empty structure (for example, there is an empty poset), that class can only be an elementary class in first-order logic if empty domains are permitted or the empty structure is removed from the class.

There are several difficulties with empty domains, however:

- Many common rules of inference are only valid when the domain of discourse is required to be nonempty. One example is the rule stating that $\phi \lor \exists x \psi$ implies $\exists x(\phi \lor \psi)$ when x is not a free variable in ϕ. This rule, which is used to put formulas into prenex normal form, is sound in nonempty domains, but unsound if the empty domain is permitted.
- The definition of truth in an interpretation that uses a variable assignment function cannot work with empty domains, because there are no variable assignment functions whose range is empty. (Similarly, one cannot assign interpretations to

constant symbols.) This truth definition requires that one must select a variable assignment function (μ above) before truth values for even atomic formulas can be defined. Then the truth value of a sentence is defined to be its truth value under any variable assignment, and it is proved that this truth value does not depend on which assignment is chosen. This technique does not work if there are no assignment functions at all; it must be changed to accommodate empty domains.

Thus, when the empty domain is permitted, it must often be treated as a special case.

Deductive Systems

A deductive system is used to demonstrate, on a purely syntactic basis, that one formula is a logical consequence of another formula. There are many such systems for first-order logic, including Hilbert-style deductive systems, natural deduction, the sequent calculus, the tableaux method, and resolution. These share the common property that a deduction is a finite syntactic object; the format of this object, and the way it is constructed, vary widely. These finite deductions themselves are often called derivations in proof theory. They are also often called proofs, but are completely formalized unlike natural-language mathematical proofs.

A deductive system is sound if any formula that can be derived in the system is logically valid. Conversely, a deductive system is complete if every logically valid formula is derivable. All of the systems discussed in this article are both sound and complete. They also share the property that it is possible to effectively verify that a purportedly valid deduction is actually a deduction; such deduction systems are called effective.

A key property of deductive systems is that they are purely syntactic, so that derivations can be verified without considering any interpretation. Thus a sound argument is correct in every possible interpretation of the language, regardless whether that interpretation is about mathematics, economics, or some other area.

In general, logical consequence in first-order logic is only semidecidable: if a sentence A logically implies a sentence B then this can be discovered (for example, by searching for a proof until one is found, using some effective, sound, complete proof system). However, if A does not logically imply B, this does not mean that A logically implies the negation of B. There is no effective procedure that, given formulas A and B, always correctly decides whether A logically implies B.

Rules of Inference

A rule of inference states that, given a particular formula (or set of formulas) with a certain property as a hypothesis, another specific formula (or set of formulas) can be derived as a conclusion. The rule is sound (or truth-preserving) if it preserves validity in the sense that whenever any interpretation satisfies the hypothesis, that interpretation also satisfies the conclusion.

For example, one common rule of inference is the rule of substitution. If t is a term and φ is a formula possibly containing the variable x, then $\varphi[t/x]$ is the result of replacing all free instances of x by t in φ. The substitution rule states that for any φ and any term t, one can conclude $\varphi[t/x]$ from φ provided that no free variable of t becomes bound during the substitution process. (If some free variable of t becomes bound, then to substitute t for x it is first necessary to change the bound variables of φ to differ from the free variables of t.)

To see why the restriction on bound variables is necessary, consider the logically valid formula φ given by $\exists x(x = y)$, in the signature of (0, 1, +, ×, =) of arithmetic. If t is the term "x + 1", the formula $\varphi[t/y]$ is $\exists x(x = x + 1)$, which will be false in many interpretations. The problem is that the free variable x of t became bound during the substitution. The intended replacement can be obtained by renaming the bound variable x of φ to something else, say z, so that the formula after substitution is $\exists z(z = x + 1)$, which is again logically valid.

The substitution rule demonstrates several common aspects of rules of inference. It is entirely syntactical; one can tell whether it was correctly applied without appeal to any interpretation. It has (syntactically defined) limitations on when it can be applied, which must be respected to preserve the correctness of derivations. Moreover, as is often the case, these limitations are necessary because of interactions between free and bound variables that occur during syntactic manipulations of the formulas involved in the inference rule.

Hilbert-style Systems and Natural Deduction

A deduction in a Hilbert-style deductive system is a list of formulas, each of which is a logical axiom, a hypothesis that has been assumed for the derivation at hand, or follows from previous formulas via a rule of inference. The logical axioms consist of several axiom schemas of logically valid formulas; these encompass a significant amount of propositional logic. The rules of inference enable the manipulation of quantifiers. Typical Hilbert-style systems have a small number of rules of inference, along with several infinite schemas of logical axioms. It is common to have only modus ponens and universal generalization as rules of inference.

Natural deduction systems resemble Hilbert-style systems in that a deduction is a finite list of formulas. However, natural deduction systems have no logical axioms; they compensate by adding additional rules of inference that can be used to manipulate the logical connectives in formulas in the proof.

Sequent Calculus

The sequent calculus was developed to study the properties of natural deduction systems. Instead of working with one formula at a time, it uses sequents, which are expressions of the form:

$$A_1,\ldots,A_n \vdash B_1,\ldots,B_k,$$

where $A_1, \ldots, A_n, B_1, \ldots, B_k$ are formulas and the turnstile symbol $\vdash$ is used as punctuation to separate the two halves. Intuitively, a sequent expresses the idea that $(A_1 \wedge \cdots \wedge A_n)$ implies $(B_1 \vee \cdots \vee B_k)$.

Tableaux Method

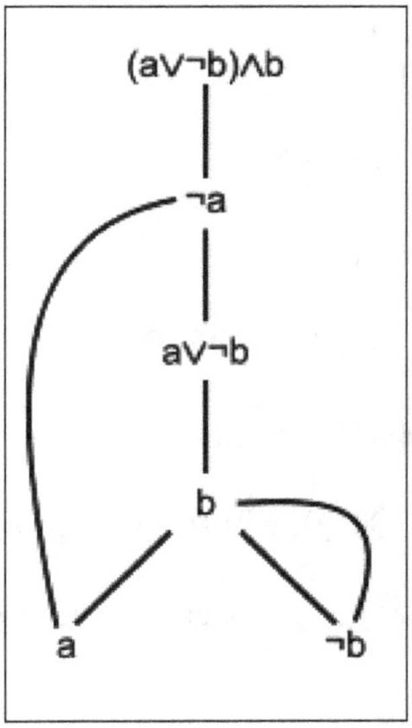

A tableaux proof for the propositional formula $((a \vee \neg b) \wedge b) \rightarrow a$.

Unlike the methods just described, the derivations in the tableaux method are not lists of formulas. Instead, a derivation is a tree of formulas. To show that a formula A is provable, the tableaux method attempts to demonstrate that the negation of A is unsatisfiable. The tree of the derivation has $\neg A$ at its root; the tree branches in a way that reflects the structure of the formula. For example, to show that $C \vee D$ is unsatisfiable requires showing that C and D are each unsatisfiable; this corresponds to a branching point in the tree with parent $C \vee D$ and children C and D.

Resolution

The resolution rule is a single rule of inference that, together with unification, is sound and complete for first-order logic. As with the tableaux method, a formula is proved by showing that the negation of the formula is unsatisfiable. Resolution is commonly used in automated theorem proving.

The resolution method works only with formulas that are disjunctions of atomic formulas; arbitrary formulas must first be converted to this form through Skolemization. The resolution rule states that from the hypotheses $A_1 \vee \cdots \vee A_k \vee C$ and $B_1 \vee \cdots \vee B_l \vee \neg C$, the conclusion $A_1 \vee \cdots \vee A_k \vee B_1 \vee \cdots \vee B_l$ can be obtained.

Provable Identities

Many identities can be proved, which establish equivalences between particular formulas. These identities allow for rearranging formulas by moving quantifiers across other connectives, and are useful for putting formulas in prenex normal form. Some provable identities include:

$$\neg \forall x P(x) \Leftrightarrow \exists x \neg P(x)$$

$$\neg \exists x P(x) \Leftrightarrow \forall x \neg P(x)$$

$$\forall x \forall y P(x,y) \Leftrightarrow \forall y \forall x P(x,y)$$

$$\exists x \exists y P(x,y) \Leftrightarrow \exists y \exists x P(x,y)$$

$$\forall x P(x) \wedge \forall x Q(x) \Leftrightarrow \forall x (P(x) \wedge Q(x))$$

$$\exists x P(x) \vee \exists x Q(x) \Leftrightarrow \exists x (P(x) \vee Q(x))$$

$$P \wedge \exists x Q(x) \Leftrightarrow \exists x (P \wedge Q(x)) \text{(where x must not occur free in P)}$$

$$P \vee \forall x Q(x) \Leftrightarrow \forall x (P \vee Q(x)) \text{ (where x must not occur free in P)}$$

Equality and its Axioms

There are several different conventions for using equality (or identity) in first-order logic. The most common convention, known as first-order logic with equality, includes the equality symbol as a primitive logical symbol which is always interpreted as the real equality relation between members of the domain of discourse, such that the "two" given members are the same member. This approach also adds certain axioms about equality to the deductive system employed. These equality axioms are:

- Reflexivity: For each variable x, $x = x$.
- Substitution for functions: For all variables x and y, and any function symbol f,

  $$x = y \rightarrow f(\ldots,x,\ldots) = f(\ldots,y,\ldots).$$

- Substitution for formulas: For any variables x and y and any formula $\varphi(x)$, if φ' is obtained by replacing any number of free occurrences of x in φ with y, such that these remain free occurrences of y, then:

  $$x = y \rightarrow (\varphi \rightarrow \varphi').$$

These are axiom schemas, each of which specifies an infinite set of axioms. The third schema is known as Leibniz's law, "the principle of substitutivity", "the indiscernibility of identicals", or "the replacement property". The second schema, involving the function symbol *f*, is (equivalent to) a special case of the third schema, using the formula:

$$x=y \rightarrow \left(f(\ldots,x,\ldots) = z \rightarrow f(\ldots,y,\ldots) = z\right).$$

Many other properties of equality are consequences of the axioms above, for example:

- Symmetry: If $x = y$ then $y = x$.
- Transitivity: If $x = y$ and $y = z$ then $x = z$.

First-order Logic without Equality

An alternate approach considers the equality relation to be a non-logical symbol. This convention is known as first-order logic without equality. If an equality relation is included in the signature, the axioms of equality must now be added to the theories under consideration, if desired, instead of being considered rules of logic. The main difference between this method and first-order logic with equality is that an interpretation may now interpret two distinct individuals as "equal" (although, by Leibniz's law, these will satisfy exactly the same formulas under any interpretation). That is, the equality relation may now be interpreted by an arbitrary equivalence relation on the domain of discourse that is congruent with respect to the functions and relations of the interpretation.

When this second convention is followed, the term normal model is used to refer to an interpretation where no distinct individuals a and b satisfy $a = b$. In first-order logic with equality, only normal models are considered, and so there is no term for a model other than a normal model. When first-order logic without equality is studied, it is necessary to amend the statements of results such as the Löwenheim–Skolem theorem so that only normal models are considered.

First-order logic without equality is often employed in the context of second-order arithmetic and other higher-order theories of arithmetic, where the equality relation between sets of natural numbers is usually omitted.

Defining Equality within a Theory

If a theory has a binary formula $A(x,y)$ which satisfies reflexivity and Leibniz's law, the theory is said to have equality, or to be a theory with equality. The theory may not have all instances of the above schemas as axioms, but rather as derivable theorems. For example, in theories with no function symbols and a finite number of relations, it is possible to define equality in terms of the relations, by defining the two terms s and t to be equal if any relation is unchanged by changing s to t in any argument.

Some theories allow other *ad hoc* definitions of equality:

- In the theory of partial orders with one relation symbol ≤, one could define $s = t$ to be an abbreviation for $s \leq t \wedge t \leq s$.
- In set theory with one relation ∈, one may define $s = t$ to be an abbreviation for $\forall x\, (s \in x \leftrightarrow t \in x) \wedge \forall x\, (x \in s \leftrightarrow x \in t)$. This definition of equality then automatically satisfies the axioms for equality. In this case, one should replace the usual axiom of extensionality, which can be stated as $\forall x \forall y [\forall z (z \in x \Leftrightarrow z \in y) \Rightarrow x = y]$, with an alternative formulation $\forall x \forall y [\forall z (z \in x \Leftrightarrow z \in y) \Rightarrow \forall z (x \in z \Leftrightarrow y \in z)]$, which says that if sets x and y have the same elements, then they also belong to the same sets.

Metalogical Properties

One motivation for the use of first-order logic, rather than higher-order logic, is that first-order logic has many metalogical properties that stronger logics do not have. These results concern general properties of first-order logic itself, rather than properties of individual theories. They provide fundamental tools for the construction of models of first-order theories.

Completeness and Undecidability

Gödel's completeness theorem, proved by Kurt Gödel in 1929, establishes that there are sound, complete, effective deductive systems for first-order logic, and thus the first-order logical consequence relation is captured by finite provability. Naively, the statement that a formula φ logically implies a formula ψ depends on every model of φ; these models will in general be of arbitrarily large cardinality, and so logical consequence cannot be effectively verified by checking every model. However, it is possible to enumerate all finite derivations and search for a derivation of ψ from φ. If ψ is logically implied by φ, such a derivation will eventually be found. Thus first-order logical consequence is semidecidable: it is possible to make an effective enumeration of all pairs of sentences (φ,ψ) such that ψ is a logical consequence of φ.

Unlike propositional logic, first-order logic is undecidable (although semidecidable), provided that the language has at least one predicate of arity at least 2 (other than equality). This means that there is no decision procedure that determines whether arbitrary formulas are logically valid. This result was established independently by Alonzo Church and Alan Turing in 1936 and 1937, respectively, giving a negative answer to the Entscheidungsproblem posed by David Hilbert and Wilhelm Ackermann in 1928. Their proofs demonstrate a connection between the unsolvability of the decision problem for first-order logic and the unsolvability of the halting problem.

There are systems weaker than full first-order logic for which the logical consequence relation is decidable. These include propositional logic and monadic predicate logic, which is first-order logic restricted to unary predicate symbols and no function symbols. Other logics with no function symbols which are decidable are the guarded fragment of first-order logic, as well as two-variable logic. The Bernays–Schönfinkel class of first-order formulas is also decidable. Decidable subsets of first-order logic are also studied in the framework of description logics.

The Löwenheim–Skolem Theorem

The Löwenheim–Skolem theorem shows that if a first-order theory of cardinality λ has an infinite model, then it has models of every infinite cardinality greater than or equal to λ. One of the earliest results in model theory, it implies that it is not possible to characterize countability or uncountability in a first-order language with a countable signature. That is, there is no first-order formula $\varphi(x)$ such that an arbitrary structure M satisfies φ if and only if the domain of discourse of M is countable (or, in the second case, uncountable).

The Löwenheim–Skolem theorem implies that infinite structures cannot be categorically axiomatized in first-order logic. For example, there is no first-order theory whose only model is the real line: any first-order theory with an infinite model also has a model of cardinality larger than the continuum. Since the real line is infinite, any theory satisfied by the real line is also satisfied by some nonstandard models. When the Löwenheim–Skolem theorem is applied to first-order set theories, the nonintuitive consequences are known as Skolem's paradox.

Compactness Theorem

The compactness theorem states that a set of first-order sentences has a model if and only if every finite subset of it has a model. This implies that if a formula is a logical consequence of an infinite set of first-order axioms, then it is a logical consequence of some finite number of those axioms. This theorem was proved first by Kurt Gödel as a consequence of the completeness theorem, but many additional proofs have been obtained over time. It is a central tool in model theory, providing a fundamental method for constructing models.

The compactness theorem has a limiting effect on which collections of first-order structures are elementary classes. For example, the compactness theorem implies that any theory that has arbitrarily large finite models has an infinite model. Thus the class of all finite graphs is not an elementary class (the same holds for many other algebraic structures).

There are also more subtle limitations of first-order logic that are implied by the compactness theorem. For example, in computer science, many situations can be modeled as a directed graph of states (nodes) and connections (directed edges). Validating such a system may require showing that no "bad" state can be reached from any "good" state. Thus one seeks to determine if the good and bad states are in different connected components of the graph. However, the compactness theorem can be used to show that connected graphs are not an elementary class in first-order logic, and there is no formula $\varphi(x,y)$ of first-order logic, in the logic of graphs, that expresses the idea that there is a path from x to y. Connectedness can be expressed in second-order logic, however, but not with only existential set quantifiers, as Σ_1^1 also enjoys compactness.

Lindström's Theorem

Per Lindström showed that the metalogical properties just discussed actually characterize first-order logic in the sense that no stronger logic can also have those properties. Lindström defined a class of abstract logical systems, and a rigorous definition of the relative strength of a member of this class. He established two theorems for systems of this type:

- A logical system satisfying Lindström's definition that contains first-order logic and satisfies both the Löwenheim–Skolem theorem and the compactness theorem must be equivalent to first-order logic.
- A logical system satisfying Lindström's definition that has a semidecidable logical consequence relation and satisfies the Löwenheim–Skolem theorem must be equivalent to first-order logic.

Limitations

Although first-order logic is sufficient for formalizing much of mathematics, and is commonly used in computer science and other fields, it has certain limitations. These include limitations on its expressiveness and limitations of the fragments of natural languages that it can describe.

For instance, first-order logic is undecidable, meaning a sound, complete and terminating decision algorithm for provability is impossible. This has led to the study of interesting decidable fragments such as C_2, first-order logic with two variables and the counting quantifiers $\exists^{\geq n}$ and $\exists^{\leq n}$ (these quantifiers are, respectively, "there exists at least n" and "there exists at most n").

Expressiveness

The Löwenheim–Skolem theorem shows that if a first-order theory has any infinite model, then it has infinite models of every cardinality. In particular, no first-order theory with an infinite model can be categorical. Thus there is no first-order theory whose only model has the set of natural numbers as its domain, or whose only model has the set of real numbers as its domain. Many extensions of first-order logic, including infinitary logics and higher-order logics, are more expressive in the sense that they do permit categorical axiomatizations of the natural numbers or real numbers. This expressiveness comes at a metalogical cost, however: by Lindström's theorem, the compactness theorem and the downward Löwenheim–Skolem theorem cannot hold in any logic stronger than first-order.

Formalizing Natural Languages

First-order logic is able to formalize many simple quantifier constructions in natural language, such as "every person who lives in Perth lives in Australia". But there are many more complicated features of natural language that cannot be expressed in (single-sorted) first-order logic. "Any logical system which is appropriate as an instrument for the analysis of natural language needs a much richer structure than first-order predicate logic".

Type	Example	Comment
Quantification over properties	If John is self-satisfied, then there is at least one thing he has in common with Peter.	Example requires a quantifier over predicates, which cannot be implemented in single-sorted first-order logic: $Zj \rightarrow \exists X(Xj \land Xp)$.
Quantification over properties	Santa Claus has all the attributes of a sadist.	Example requires quantifiers over predicates, which cannot be implemented in single-sorted first-order logic: $\forall X(\forall x(Sx \rightarrow Xx) \rightarrow Xs)$.
Predicate adverbial	John is walking quickly.	Example cannot be analysed as $Wj \land Qj$; predicate adverbials are not the same kind of thing as second-order predicates such as colour.
Relative adjective	Jumbo is a small elephant.	Example cannot be analysed as $Sj \land Ej$; predicate adjectives are not the same kind of thing as second-order predicates such as colour.
Predicate adverbial modifier	John is walking very quickly.	-
Relative adjective modifier	Jumbo is terribly small.	An expression such as "terribly", when applied to a relative adjective such as "small", results in a new composite relative adjective "terribly small".
Prepositions	Mary is sitting next to John.	The preposition "next to" when applied to "John" results in the predicate adverbial "next to John".

Many-sorted Logic

Many-sorted logic can reflect formally our intention not to handle the universe as a homogeneous collection of objects, but to partition it in a way that is similar to types in typeful programming. Both functional and assertive "parts of speech" in the language of the logic reflect this typeful partitioning of the universe, even on the syntax level: Substitution and argument passing can be done only accordingly, respecting the "sorts".

There are various ways to formalize the intention mentioned above; a *many-sorted logic* is any package of information which fulfills it. In most cases, the following are given:

- A set of sorts, S.

- An appropriate generalization of the notion of *signature* to be able to handle the additional information that comes with the sorts.

The domain of discourse of any structure of that signature is then fragmented into disjoint subsets, one for every sort.

Example:

When reasoning about biological organisms, it is useful to distinguish two sorts: plant and animal. While a function $\text{mother} : \text{animal} \to \text{animal}$ makes sense, a similar function $\text{mother} : \text{plant} \to \text{plant}$ usually does not. Many-sorted logic allows one to have terms like mother(lassie), but to discard terms like mother(my_favorite_oak) as syntactically ill-formed.

Order-sorted Logic

While *many-sorted* logic requires two distinct sorts to have disjoint universe sets, *order-sorted* logic allows one sort s_1 to be declared a subsort of another sort , usually by writing $s_1 \subseteq s_2$ or similar syntax. In the above example, it is desirable to declare:

- dog $\subseteq$ carnivore.
- dog $\subseteq$ mammal.
- carnivore $\subseteq$ animal.
- mammal $\subseteq$ animal.
- animal $\subseteq$ organism.
- plant $\subseteq$ organism and so on.

Wherever a term of some sort s is required, a term of any subsort of s may be supplied instead. For example, assuming a function declaration $\text{mother} : \text{animal} \to \text{animal}$, and a constant declaration $\text{lassie} : \text{dog}$, the term mother(lassie) is perfectly valid and has the sort animal. In order to supply the information that the mother of a dog is a dog in turn, another declaration $\text{mother} : \text{dog} \to \text{dog}$ may be issued; this is called *function overloading*, similar to overloading in programming languages.

Order-sorted logic can be translated into unsorted logic, using a unary predicate $p_i(x)$ for each sort s_i, and an axiom $\forall x(p_i(x) \to p_j(x))$ for each subsort declaration $s_i \subseteq s_j$. The reverse approach was successful in automated theorem proving: in 1985, Christoph Walther could solve a then benchmark problem by translating it into order-sorted logic, thereby boiling it down an order of magnitude, as many unary predicates turned into sorts.

In order to incorporate order-sorted logic into a clause-based automated theorem prover, a corresponding *order-sorted unification* algorithm is necessary, which requires for any two declared sorts s_1, s_2 their intersection $s_1 \cap s_2$ to be declared, too: if x_1 and x_2 are variables of sort s_1 and s_2, respectively, the equation $x_1 \stackrel{?}{=} x_2$ has the solution $\{x_1 = x, x_2 = x\}$, where $x : s_1 \cap s_2$.

Smolka generalized order-sorted logic to allow for parametric polymorphism. In his framework, subsort declarations are propagated to complex type expressions. As a programming example, a parametric sort $\text{list}(X)$ may be declared (with X being a type parameter as in a C++ template), and from a subsort declaration $\text{int} \subseteq \text{float}$ the relation $\text{list}(\text{int}) \subseteq \text{list}(\text{float})$ is automatically inferred, meaning that each list of integers is also a list of floats.

Schmidt-Schauß generalized order-sorted logic to allow for term declarations. As an example, assuming subsort declarations $\text{even} \subseteq \text{int}$ and $\text{odd} \subseteq \text{int}$, a term declaration like $\forall i : \text{int}. (i + i) : \text{even}$ allows to declare a property of integer addition that could not be expressed by ordinary overloading.

Infinitary Logic

An infinitary logic is a logic that allows infinitely long statements and infinitely long proofs. Some infinitary logics may have different properties from those of standard first-order logic. In particular, infinitary logics may fail to be compact or complete. Notions of compactness and completeness that are equivalent in finitary logic sometimes are not so in infinitary logics. Therefore for infinitary logics, notions of strong compactness and strong completeness are defined.

Considering whether a certain infinitary logic named Ω-logic is complete promises to throw light on the continuum hypothesis.

A Word on Notation and the Axiom of Choice

As a language with infinitely long formulae is being presented, it is not possible to write such formulae down explicitly. To get around this problem a number of notational conveniences, which, strictly speaking, are not part of the formal language, are used. $\cdots$ is used to point out an expression that is infinitely long. Where it is unclear, the length of the sequence is noted afterwards. Where this notation becomes ambiguous or confusing, suffixes such as $\vee_{\gamma<\delta} A_\gamma$ are used to indicate an infinite disjunction over a set of formulae of cardinality δ. The same notation may be applied to quantifiers for example $\forall_{\gamma<\delta} V_\gamma :$. This is meant to represent an infinite sequence of quantifiers for each V_γ where $\gamma < \delta$.

All usage of suffixes and $\cdots$ are not part of formal infinitary languages.

The axiom of choice is assumed (as is often done when discussing infinitary logic) as this is necessary to have sensible distributivity laws.

Definition of Hilbert-type Infinitary Logics

A first-order infinitary logic $L_{\alpha,\beta}$, α regular, $\beta = 0$ or $\omega \leq \beta \leq \alpha$, has the same set of symbols as a finitary logic and may use all the rules for formation of formulae of a finitary logic together with some additional ones:

- Given a set of formulae $A = \{A_\gamma \mid \gamma < \delta < \alpha\}$ then $(A_0 \vee A_1 \vee \cdots)$ and $(A_0 \wedge A_1 \wedge \cdots)$ are formulae. (In each case the sequence has length δ.)
- Given a set of variables $V = \{V_\gamma \mid \gamma < \delta < \beta\}$ and a formula A_0 then $\forall V_0 : \forall V_1 \cdots (A_0)$ and $\exists V_0 : \exists V_1 \cdots (A_0)$ are formulae. (In each case the sequence of quantifiers has length δ.)

The concepts of free and bound variables apply in the same manner to infinite formulae. Just as in finitary logic, a formula all of whose variables are bound is referred to as a *sentence.*

A theory T in infinitary logic $L_{\alpha,\beta}$ is a set of sentences in the logic. A proof in infinitary logic from a theory T is a sequence of statements of length γ which obeys the following conditions: Each statement is either a logical axiom, an element of T, or is deduced from previous statements using a rule of inference. As before, all rules of inference in finitary logic can be used, together with an additional one:

- Given a set of statements $A = \{A_\gamma \mid \gamma < \delta < \alpha\}$ which have occurred previously in the proof then the statement $\wedge_{\gamma<\delta} A_\gamma$ can be inferred.

The logical axiom schemata specific to infinitary logic are presented below. Global schemata variables: δ and γ such that $0 < \delta < \alpha$.

- $((\wedge_{\epsilon<\delta}(A_\delta \Rightarrow A_\epsilon)) \Rightarrow (A_\delta \Rightarrow \wedge_{\epsilon<\delta} A_\epsilon))$.
- For each $\gamma < \delta$, $\left((\wedge_{\epsilon<\delta} A_\epsilon) \Rightarrow A_\gamma\right)$.
- Chang's distributivity laws (for each γ): $(\vee_{\mu<\gamma}(\wedge_{\delta<\gamma} A_{\mu,\delta}))$, where $\forall\mu\forall\delta\exists\epsilon < \gamma : A_{\mu,\delta} = A_\epsilon$ or $A_{\mu,\delta} = \neg A_\epsilon A_{\mu,\delta} = \neg A_\epsilon$, and $\forall g \in \gamma^\gamma \exists \epsilon < \gamma : \{A_\epsilon, \neg A_\epsilon\} \subseteq \{A_{\mu,g(\mu)} : \mu < \gamma\}$.
- For $\gamma < \alpha$, $((\wedge_{\mu<\gamma}(\vee_{\delta<\gamma} A_{\mu,\delta})) \Rightarrow (\vee_{\epsilon<\gamma^\gamma}(\wedge_{\mu<\gamma} A_{\mu,\gamma_\epsilon(\mu)})))$, where $\{\gamma_\epsilon : \epsilon < \gamma^\gamma\}$ is a well ordering of γ^γ.
- The last two axiom schemata require the axiom of choice because certain sets must be well orderable. The last axiom schema is strictly speaking unnecessary as Chang's distributivity laws imply it, however it is included as a natural way to allow natural weakenings to the logic.

Completeness, Compactness and Strong Completeness

A theory is any set of statements. The truth of statements in models are defined by recursion and will agree with the definition for finitary logic where both are defined. Given a theory T a statement is said to be valid for the theory T if it is true in all models of T.

A logic $L_{\alpha,\beta}$ is complete if for every sentence S valid in every model there exists a proof of S. It is strongly complete if for any theory T for every sentence S valid in T there is a proof of S from T. An infinitary logic can be complete without being strongly complete.

A cardinal $\kappa \neq \omega$ is weakly compact when for every theory T in $L_{\kappa,\kappa}$ containing at most κ many formulas, if every $S \subseteq T$ of cardinality less than κ has a model, then T has a model. A cardinal $\kappa \neq \omega$ is strongly compact when for every theory T in $L_{\kappa,\kappa}$, without restriction on size, if every $S \subseteq T$ of cardinality less than κ has a model, then T has a model.

Concepts Expressible in Infinitary Logic

In the language of set theory the following statement expresses foundation:

$$\forall_{\gamma<\omega} V_{\gamma} : \neg \wedge_{\gamma<\omega} V_{\gamma+} \in V_{\gamma}$$

Unlike the axiom of foundation, this statement admits no non-standard interpretations. The concept of well-foundedness can only be expressed in a logic which allows infinitely many quantifiers in an individual statement. As a consequence many theories, including Peano arithmetic, which cannot be properly axiomatised in finitary logic, can be in a suitable infinitary logic. Other examples include the theories of non-archimedean fields and torsion-free groups. These three theories can be defined without the use of infinite quantification; only infinite junctions are needed.

Complete Infinitary Logics

Two infinitary logics stand out in their completeness. These are $L_{\omega,\omega}$ and $L_{\omega_1,\omega}$. The former is standard finitary first-order logic and the latter is an infinitary logic that only allows statements of countable size.

$L_{\omega,\omega}$ is also strongly complete, compact and strongly compact.

$L_{\omega_1,\omega}$ fails to be compact, but it is complete (under the axioms given above). Moreover, it satisfies a variant of the Craig interpolation property.

If $L_{\alpha,\alpha}$ is strongly complete (under the axioms given above) then α is strongly compact (because proofs in these logics cannot use α or more of the given axioms).

Higher-order Logic

In mathematics and logic, a higher-order logic is a form of predicate logic that is distinguished from first-order logic by additional quantifiers and, sometimes, stronger semantics. Higher-order logics with their standard semantics are more expressive, but their model-theoretic properties are less well-behaved than those of first-order logic.

The term "higher-order logic", abbreviated as HOL, is commonly used to mean higher-order simple predicate logic. Here "simple" indicates that the underlying type theory is the theory of simple types, also called the simple theory of types. Leon Chwistek and Frank P. Ramsey proposed this as a simplification of the complicated and clumsy ramified theory of types specified in the *Principia Mathematica* by Alfred North Whitehead and Bertrand Russell. Simple types is nowadays sometimes also meant to exclude polymorphic and dependent types.

Quantification Scope

First-order logic quantifies only variables that range over individuals; second-order logic, in addition, also quantifies over sets; third-order logic also quantifies over sets of sets, and so on.

Higher-order logic is the union of first-, second-, third-, ..., *n*th-order logic; *i.e.*, higher-order logic admits quantification over sets that are nested arbitrarily deeply.

Semantics

There are two possible semantics for higher order logic.

In the standard or full semantics, quantifiers over higher-type objects range over *all* possible objects of that type. For example, a quantifier over sets of individuals ranges over the entire powerset of the set of individuals. Thus, in standard semantics, once the set of individuals is specified, this is enough to specify all the quantifiers. HOL with standard semantics is more expressive than first-order logic. For example, HOL admits categorical axiomatizations of the natural numbers, and of the real numbers, which are impossible with first-order logic. However, by a result of Kurt Gödel, HOL with standard semantics does not admit an effective, sound, and complete proof calculus.The model-theoretic properties of HOL with standard semantics are also more complex than those of first-order logic. For example, the Löwenheim number of second-order logic is already larger than the first measurable cardinal, if such a cardinal exists. The Löwenheim number of first-order logic, in contrast, is $\aleph_0$, the smallest infinite cardinal.

In Henkin semantics, a separate domain is included in each interpretation for each higher-order type. Thus, for example, quantifiers over sets of individuals may range over only a subset of the powerset of the set of individuals. HOL with these semantics is equivalent to many-sorted first-order logic, rather than being stronger than first-order logic. In particular, HOL with Henkin semantics has all the model-theoretic properties of first-order logic, and has a complete, sound, effective proof system inherited from first-order logic.

Examples and Properties

Higher order logics include the offshoots of Church's Simple theory of types and the various forms of intuitionistic type theory. Gérard Huet has shown that unifiability is undecidable in a type theoretic flavor of third-order logic, that is, there can be no algorithm to decide whether an arbitrary equation between third-order (let alone arbitrary higher-order) terms has a solution.

Up to a certain notion of isomorphism, the powerset operation is definable in second-order logic. Using this observation, Jaakko Hintikka established in 1955 that second-order logic can simulate higher-order logics in the sense that for every formula of a higher order-logic one can find an equisatisfiable formula for it in second-order logic.

The term "higher-order logic" is assumed in some context to refer to *classical* higher-order logic. However, modal higher-order logic has been studied as well. According to several logicians, Gödel›s ontological proof is best studied (from a technical perspective) in such a context.

References

- Formal-system, topic: britannica.com, Retrieved 24 March, 2020
- Carlos Caleiro, Ricardo Gonçalves (2006). "On the algebraization of many-sorted logics". Proc. 18th int. conf. on Recent trends in algebraic development techniques (WADT) (PDF). Springer. pp. 21–36. ISBN 978-3-540-71997-7
- Formal- logic, topic: britannica.com, Retrieved 13 June, 2020
- Rautenberg, Wolfgang (2010), A Concise Introduction to Mathematical Logic (3rd ed.), New York, NY: Springer Science+Business Media, doi:10.1007/978-1-4419-1221-3, ISBN 978-1-4419-1220-6
- Algebraic-logic: encyclopediaofmath.org, Retrieved 25 March, 2020

Permissions

All chapters in this book are published with permission under the Creative Commons Attribution Share Alike License or equivalent. Every chapter published in this book has been scrutinized by our experts. Their significance has been extensively debated. The topics covered herein carry significant information for a comprehensive understanding. They may even be implemented as practical applications or may be referred to as a beginning point for further studies.

We would like to thank the editorial team for lending their expertise to make the book truly unique. They have played a crucial role in the development of this book. Without their invaluable contributions this book wouldn't have been possible. They have made vital efforts to compile up to date information on the varied aspects of this subject to make this book a valuable addition to the collection of many professionals and students.

This book was conceptualized with the vision of imparting up-to-date and integrated information in this field. To ensure the same, a matchless editorial board was set up. Every individual on the board went through rigorous rounds of assessment to prove their worth. After which they invested a large part of their time researching and compiling the most relevant data for our readers.

The editorial board has been involved in producing this book since its inception. They have spent rigorous hours researching and exploring the diverse topics which have resulted in the successful publishing of this book. They have passed on their knowledge of decades through this book. To expedite this challenging task, the publisher supported the team at every step. A small team of assistant editors was also appointed to further simplify the editing procedure and attain best results for the readers.

Apart from the editorial board, the designing team has also invested a significant amount of their time in understanding the subject and creating the most relevant covers. They scrutinized every image to scout for the most suitable representation of the subject and create an appropriate cover for the book.

The publishing team has been an ardent support to the editorial, designing and production team. Their endless efforts to recruit the best for this project, has resulted in the accomplishment of this book. They are a veteran in the field of academics and their pool of knowledge is as vast as their experience in printing. Their expertise and guidance has proved useful at every step. Their uncompromising quality standards have made this book an exceptional effort. Their encouragement from time to time has been an inspiration for everyone.

The publisher and the editorial board hope that this book will prove to be a valuable piece of knowledge for students, practitioners and scholars across the globe.

Index

www.ingramcontent.com/pod-product-compliance
Lightning Source LLC
LaVergne TN
LVHW080619210926
848485LV00003B/186